Delivering Change

Towards fit-for-purpose governance of adaptation to flooding and drought

Cover:

Delivering change. The de-poldering of the Noordwaard represents a new way of flood risk management in the Netherlands in which flood risks are reduced by creating space for discharging peak river flows (where appropriate). Source: https://beeldbank.rws.nl, Rijkswaterstaat , Ruimte voor de Rivier / Con Mönnich

Delivering Change

Towards fit-for-purpose governance of adaptation to flooding and drought

Dissertation

Submitted in fulfillment of the requirements of
the Board for Doctorates of Delft University of Technology
and of the Academic Board of the UNESCO-IHE
Institute for Water Education
for the Degree of DOCTOR
to be defended in public on
Friday, 14 February 2014 at 12:30 hours
in Delft, The Netherlands

by

Jeroen Sebastiaan RIJKE

Master of Science in Civil Engineering
born in Rotterdam, the Netherlands

This dissertation has been approved by the supervisor:
Prof. dr.ir. C. Zevenbergen

Composition of Doctoral Committee:

Chairman	Rector Magnificus TU Delft
Vice-chairman	Rector UNESCO-IHE
Prof. dr.ir. C. Zevenbergen	UNESCO-IHE / TU Delft, supervisor
Prof. dr. R.R. Brown	Monash University, Australia
Prof. R.M. Ashley	University of Sheffield, United Kingdom
Prof. dr. J. Edelenbos	Erasmus University Rotterdam
Prof. dr. ir. M.J.C.M. Hertogh	TU Delft
Ir. I. de Boer	Rijkswaterstaat, Utrecht
Prof. dr. ing. S. Schaap	TU Delft, reserve member

CRC Press/Balkema is an imprint of the Taylor & Francis Group, an informa business

Published by:
CRC Press/Balkema
PO Box 11320, 2301 EH Leiden, The Netherlands
e-mail: Pub.NL@taylorandfrancis.com
www.crcpress.com – www.taylorandfrancis.com

ISBN 978-1-138-02633-9 (Taylor & Francis Group)

Preface

This dissertation combines the outcomes of two research projects that were conducted between September 2009 and June 2013.

The first project (September 2009 – March 2012) was funded by the Cities as Water Supply Catchments research programme and explored how urban water governance in Australia could enhance resilience to drought. I have collaborated in this project as a visiting researcher at the School of Geography and Environmental Science at Monash University in Melbourne.

The second project (January 2010 – June 2013) entailed a scientific evaluation of the Room for the River programme that was commissioned by Rijkswaterstaat to UNESCO-IHE. The scientific rigour of this evaluation was evaluated by a scientific advisory board that included Prof. dr. Jim Hall (University of Oxford), Prof. dr. Jurian Edelenbos (Erasmus University) and Prof. dr. Rebekah Brown (Monash University). In addition, a user panel consisting of experts with policy and advisory roles reflected on the practical relevance of the evaluation.

These two projects have resulted in several interrelated papers of which five are included in this thesis. The presented thesis should, therefore, be considered a thesis by papers. However, from the outset of the research, these papers were written to an overall plan to create a coherent story. This plan was driven by the ambition to assist policy makers and project managers in delivering adaptation action in practice. This thesis is the result of that plan that, was also adapted to opportunities that came on to my path and has evolved and deepened as I progressed.

Jeroen Rijke
February, 2014

This research is funded by:

Summary

Overcoming hurdles to adaptation

There is a great need for societies to adapt to climate change in order to anticipate increasing risks and/or seize new opportunities. The IPCC defines adaptation as *"the adjustment in natural or human systems in response to actual or expected climatic stimuli or their effects, which moderates harm or exploits beneficial opportunities"* (Parry et al., 2007, p.6). Adaptation to climate change is commonly referred to as a governance challenge. With regard to water management, the focus of this thesis, this governance challenge can be broken down into several parts. Firstly, the practical implementation of available innovative technologies and knowledge required to develop adaptive water management systems is slow. Secondly, it is nowadays frequently suggested that new modes of governance are needed that are effective under conditions of high complexity and uncertainty. These approaches would involve, for example, multiple disciplines, multiple government levels, the community, the private sector and academia. Adger and colleagues summarise these governance challenges nicely, by suggesting that *"adaptation to climate change is limited by the values, perceptions, processes and power structures within society"* (Adger et al., 2009, p.349).

Delivering adaptation action in the water sector is about delivering change, because adaptation of a water system requires a transition in the form of technological innovation and/or the adoption of new governance approaches. In this thesis, it is explained that governance approaches for the effective delivery of adaptation action to manage changes in flood and drought risks should be considered on a case by case basis. However, the analyses provide several ingredients that assist policy makers, planners and project managers in developing effective governance approaches for adaptation to flooding and drought:

1. A procedure for establishing fit-for-purpose governance reform. (chapter 2)
2. A pattern of governance approaches that are typically effective during the early, mid and late stages of transformation. (chapter 3)

3. A checklist for the availability of the required ingredients for change. (chapter 4)
4. A set of attributes for effective design and management of adaptation programmes (chapter 5)
5. Recommendations for aligning governance of strategic planning and delivery of adaptation. (chapter 6)

Fit-for-purpose governance

Drawing on an extensive, critical literature review of adaptive governance, network management and institutional analysis, I argue that the constraints to the governance of adaptation relate to a large extent to the inability of practitioners and policy makers to cope with complexity and various uncertainties: i) ambiguous purposes and objectives of what should be achieved with governance; ii) unclear contextual conditions in which governance takes place; and, iii) uncertainty around the effectiveness of different governance strategies. To address such practical challenges, I introduce a 'fit-for-purpose' framework consisting of three key ingredients for developing a diagnostic approach for making adaptive governance operational. This concept is meant to be used as an indication of the effectiveness of governance structures and processes and define it as a measure of the adequacy of the functional purposes that governance structures and processes have to fulfil at a certain point in time. In other words, are existing and proposed governance structures and processes fit for their purpose? While adaptive governance focuses on responding to (potential) change, fit-for-purpose governance is specifically considering the (future) functions that the social and physical components of a particular system, such as an urban water system, have to fulfil. As such, the fit-for-purpose governance framework provides an alternative starting point for developing the much sought-after guidance for policy and decision makers to evaluate the effectiveness of established governance arrangements and to predict the likelihood of success of institutional reform.

A pattern of effective governance during consecutive stages of transformation

Based on a comparison of governance reform of urban water management in three Australian cities, a pattern of effective governance configurations during consecutive stages of system transformation is identified. By linking the (perceived) effectiveness of centralised, decentralised, formal and informal governance approaches to the requirements of consecutive stages of respectively adaptive cycles and transition stages, it was found that different configurations of these governance approaches are needed during different stages of adapting to drought and transitioning to a water sensitive city that is resilient to immediate and gradual change. The research insights suggest that decentralised and informal governance approaches are particularly effective in early stages of transformation processes (i.e. adaptation and transition processes), whilst formal and centralised approaches become more effective during later stages of transformation (Table S.1). This pattern of effective governance configurations can be used to provide guidance for urban water governance reform to policy makers and governance evaluators.

Table S.1 Effective governance during consecutive stages of transformation

Transition stage	Adaptive cycle phase	Typical activities	Effective governance approaches
Pre-development	-	Network formation, experimentation, learning.	*Decentralised and informal*: to establish and nurture new relationships and test innovations
Take-off	Re-organisation / renewal	Response to a crisis or establishment of a policy decision.	*Hybrid*: formal policy decision to catalyse and/or coordinate activities, and informal and decentralised learning to further test innovations
Acceleration	Growth / exploitation	Increasing implementation of innovation.	*Hybrid*: centralised policy to enable activities, decentralised implementation, informal network to distribute tacit knowledge, coordinated capacity building to create synergies and avoid inefficient use of resources.
Stabilisation	Conservation	Regulation and legislation to establish the status quo	*Centralised and formal*: to adjust or establish legislative frameworks and coordinated capacity building to convince and enable laggards to adopt innovative approaches and safeguard a new status quo.
-	Collapse / release	Losing faith, searching for new/alternative solutions	*Decentralised and informal*: to search for alternative solutions and share experiences.

Criteria for change

Through application of an existing 'transitions governance' framework, a set of criteria for establishing structural system change (i.e. system-wide adaptation or a transition) is tested for the context of river flood protection in the Netherlands. This led to the conclusion that system transformation depends on the presence of eight enabling factors:

1. A narrative, metaphor and image that support a clear vision for change
2. A regulatory and compliance agenda
3. Economic justification
4. Policy and planning frameworks and institutional design
5. Leadership
6. Capacity building and demonstration
7. Public engagement and behaviour change
8. Research and partnerships with policy/practice

These criteria can be used as a checklist for policy and decision makers to establish system transformation. Factors (1) to (4) are requirements for developing and performing new practices, whilst factors (5) to (8) are needed for enabling new practices. In case of river flood protection in the Netherlands, this applied to integrating flood protection and spatial quality objectives. Insight into the absence of one or more of the eight enabling factors for system transformation gives insight as to what governance arrangements are not fit for the purpose of delivering change and, thus, why such transformational processes are hampered by inadequate governance.

Delivering adaptation projects

The Dutch case of river flood protection illustrates that a large scale infrastructure programme, can have a significant impact on how a transition towards system-wide adaptation evolves. Combining insights from the project and programme management literature with the case study findings of the €2.4 billion Room for the River flood protection programme in the Netherlands, revealed a (preliminary) set of attributes for effective programme management:

1) A *clear programme vision* that is widely supported by all relevant stakeholders.
2) A *clear priority focus* that provides opportunities to connect stakeholder ambitions to the overall programme objectives.
3) A *transparent programme planning framework* that outlines the boundary conditions and roles of the stakeholders.
4) *Programme governance* involving internal and external stakeholders that matches the vision, priority focus and planning framework of the programme to enhance the legitimacy and quality of the programme and its projects.
5) *Appropriate programme coordination* to monitor progress and management performance and, if needed, assist projects in achieving their objectives.
6) *Programme adaptation* to adjust the programme's organisation or outcomes to the context of the individual projects and the programme as a whole.

Furthermore, it was found that a combined strategic/performance focus at the level of both programme and project management that enables a collaborative approach between programme and project management. This particularly enables effective stakeholder collaboration, coordination and adaptation of the programme to contextual changes, newly acquired insights and the changing needs of consecutive planning stages, which positively contributes to the performance of the programme as a whole.

Planning for and delivery of adaptation

Previous research on governance of adaptation has focused predominantly on strategic planning for adaptation and has largely overlooked the delivery of adaptation in practice. Meanwhile, there is a gap between aspirations for adaptive water management systems and the realisation thereof. Based on a comparison of cases of adaptation in the water sectors in the Netherlands and Australia, which are globally regarded as leading in terms of implementing innovative water management approaches, the coming about of adaptation action was analysed by investigating the interactions between the governance for strategic planning and the governance for the delivery of adaptation. These cases show that governance of strategic planning can enhance delivery through creating the conditions that are needed to deliver adaptation action effectively, including stakeholder support, a broad knowledge base and an allocated investment budget for the realisation of adaptation action. Vice versa, both cases show that governance of delivery can be influential for strategic planning of new adaptation actions through knowledge and relationships that are developed for the realisation of adaptation action. Hence, it can be concluded that governance for strategic planning and governance for the delivery of adaptation action can reinforce each other. As a consequence, the governance of adaptation scholarship would benefit from refocusing its current emphasis on strategic planning towards an approach that also incorporates a lens for implementation in order to turn aspirations into reality.

Samenvatting

Het overwinnen van obstakels voor adaptatie

Toenemende klimaatgerelateerde risico's vergroten de noodzaak voor onze samenleving om adaptatiemaatregelen te nemen. Het IPCC definiëert adaptatie als *"the adjustment in natural or human systems in response to actual or expected climatic stimuli or their effects, which moderates harm or exploits beneficial opportunities"* (Parry et al., 2007, p.6). Klimaatadaptatie wordt vaak aangeduid als een governance-vraagstuk. Met betrekking tot het waterbeheer, de focus van dit proefschrift, zijn in dit verband twee constateringen relevant. Ten eerste blijft de praktische toepassing van innovatieve technologieën en kennis die benodigd zijn voor adaptief waterbeheer vaak achter bij wat er beschikbaar is. Ten tweede wordt er vaak gesuggereerd dat nieuwe vormen van governance, waarin samenwerking tussen verschillende actoren en disciplines centraal staat, nodig zijn om op effectieve wijze om te kunnen gaan met complexiteit en onzekerheid. Adger en collega's vatten deze twee uitdagingen op het gebied van governance mooi samen, door te stellen dat *"adaptation to climate change is limited by the values, perceptions, processes and power structures within society"* (Adger et al., 2009, p.349).

Het bewerkstelligen van adaptatie in de watersector gaat over het bewerkstelligen van verandering. Adaptatie van een watersysteem vereist immers een transitie in de vorm van technologische innovatie en / of nieuwe vormen van organisatie. In dit proefschrift wordt uitgelegd dat de governance-benadering voor het effectief bewerkstelligen van adaptatie ten aanzien van veranderingen in overstromings- en droogterisico's per geval apart moet worden beschouwd. Echter, de analyses in dit proefschrift bieden een aantal bouwstenen die beleidsmakers, planners en projectmanagers assisteren bij het ontwikkelen van effectieve governance-benaderingen voor adaptatie ten aanzien van overstromingen en droogte:

1. Een procedure voor het vaststellen van 'fit-for-purpose' governance (hoofdstuk 2).
2. Een patroon van governance-benaderingen die typisch geschikt zijn tijdens de vroege, midden- en late stadia van transformatie van een watersysteem (hoofdstuk 3).

3. Een checklist van de ingrediënten die nodig zijn voor het bewerkstelligen van verandering (hoofdstuk 4).
4. Een set van aanbevelingen voor effectief management van uitvoeringsprogramma's voor adaptatie (hoofdstuk 5).
5. Aanbevelingen om governance voor planning en uitvoering van adaptatie meer in overeenstemming te aanschouwen (hoofdstuk 6).

Fit-for-purpose governance

Op basis van literatuuronderzoek over adaptief management, netwerk management en institutionele analyse, beargumenteer ik dat de obstakels voor governance van adaptatie voor een groot deel te maken hebben met het onvermogen van zowel de beleidswereld als de praktijk om om te gaan met complexiteit en verschillende onzekerheden. Dit onvermogen is aanleiding geweest om in het kader van dit onderzoek een procedure te ontwikkelen om de geschiktheid, ofwel de 'fit-for-purpose', van governance te bepalen. Hierin staan drie activiteiten centraal: 1) het bepalen van de doelstelling van een toegepaste of beoogde governance benadering; 2) beschrijving van invloedrijke contextuele factoren op de werking van een governance-benadering; en 3) het vermogen van een governance-benadering om haar doel te bereiken binnen de beschreven context. Als zodanig, biedt het denken over de 'fit-for-purpose' van governance een handreiking voor beleidsmakers en besluitvormers om de doeltreffendheid van de governance-benaderingen te evalueren en om de waarschijnlijkheid van succes van de institutionele hervorming te voorspellen.

Een patroon van effectieve governance tijdens opeenvolgende stadia van transformatie

Op basis van een vergelijkend onderzoek over de opkomst van Water Sensitive Urban Design in drie Australische steden, is een patroon van governance-benaderingen geïdentificeerd die effectief zijn tijdens opeenvolgende fasen van transformatie (i.e. adaptatiecyclus of transitie). Hieruit blijkt dat de effectiviteit van een bepaalde aanpak per fase verschilt. De bevindingen suggereren dat gedecentraliseerde en informele benaderingen over het algemeen geschikt zijn tijdens de eerste stadia van transformatieprocessen, terwijl de formele en gecentraliseerde benaderingen juist effectiever blijken tijdens de latere stadia van transformatie (Tabel S.2). Deze bevindingen geven beleidsmakers en analisten inzicht in hoe hervormingen van het (stedelijk) waterbeheer op effectieve wijze gestalte kunnen krijgen.

Criteria voor verandering

Door toepassing van een bestaand 'transformative governance' raamwerk, is een set van criteria voor het bereiken van structurele systeemverandering getest in de context van hoogwaterbescherming in Nederlandse riviergebieden. Dit leidde tot de conclusie dat dergelijke transformatie afhankelijk is van de aanwezigheid van acht factoren:

1. Een verhaal, metafoor en beeldvorming ter ondersteuning van een heldere visie voor verandering
2. Adequate regelgeving en mechanismen voor handhaving
3. Economische rechtvaardiging
4. Beleidskaders, planvorminginstrumenten en institutionele kaders
5. Leiderschap
6. 'Capacity building' en demonstratie projecten
7. Publieke betrokkenheid en gedragsverandering
8. Wetenschappelijke partnerschappen met beleid / praktijk

Wanneer een of meer van de acht factoren afwezig of onderontwikkeld is, zal dit veranderingsprocessen belemmeren. De bovenstaande criteria kunnen dus door beleidsmakers en besluitvormers worden gebruikt als een checklist om transformatie te bewerkstelligen.

De realisatie van adaptatieprojecten

Het hoogwaterbeschermingsprogramma Ruimte voor de Rivier illustreert dat een grootschalige infrastructuurprogramma in Nederland een aanzienlijke uitwerking kan hebben op hoe systeembrede adaptatie evolueert. Door inzichten uit de literatuur over project en programma management te combineren met de analyse van het Ruimte voor de Rivier programma, is een set van kenmerken van effectief programma management geïdentificeerd:

1. Een duidelijke *programma visie* die breed gedragen door alle belanghebbenden.
2. Een duidelijke *prioriteitstelling van de scope* die mogelijkheden biedt om ambities belanghebbenden te verbinden met de algemene doelstellingen van het programma.
3. Een transparant *planningskader* waarin de randvoorwaarden en de rollen van de betrokkenen duidelijk vastgesteld zijn.
4. *Programma governance* ten behoeve van het betrekken van interne en externe belanghebbenden die afgestemd is met de visie, prioriteitstelling en planningskader van het programma om de legitimiteit en kwaliteit van het programma en de projecten te verbeteren.
5. Passende *coördinatiemechanismen* om de voortgang en prestaties van het management te controleren en, indien nodig, projecten te assisteren bij het bereiken van hun doelstellingen.
6. Een *adaptieve programma organisatie* die zich aan kan passen aan de context van de individuele projecten en het programma als geheel.

Bovendien laat het Ruimte voor de Rivier programma zien dat de effectiviteit van een uitvoeringsprogramma baat heeft bij een goede samenwerking tussen het programmabureau en de projectteams. Een dergelijke goede samenwerking wordt bevorderd als het management van beiden geschiedt op een wijze die zowel oog heeft voor het behalen van prestaties als de strategische aspecten, zoals het delen van kennis en het rekening houden met politieke gevoeligheden.

Planning en realisatie van adaptatie

Bestaand onderzoek over governance van adaptatie is voornamelijk gericht op de strategische (beleids-) aspecten van adaptatie, terwijl de realisatie van adaptatiemaatregelen in de praktijk grotendeels over het hoofd wordt gezien. Ondertussen is er in veel gevallen een kloof tussen de ambitie ten aanzien van adaptieve watersystemen en wat er daarvan in de praktijk wordt gerealiseerd. Gebaseerd op een vergelijking van twee toonaangevende cases van adaptatie in de watersectoren in Nederland (Ruimte voor de Rivier) en Australië (Water Sensitive Urban Design), is de totstandkoming van adaptatiemaatregelen geanalyseerd. Deze cases illustreren dat governance voor planning een positief effect kan hebben op de uitvoering van adaptatiemaatregelen door het creëren van de voorwaarden die nodig zijn om zulke maatregelen effectief te realiseren, zoals legitimiteit en steun, een brede kennisbasis, en financiering. Vice versa, blijkt uit beide cases dat governance voor realisatie van adaptatiemaatregelen en uitvoeringsprogramma's van strategische waarde kan zijn voor nieuwe adaptatiemaatregelen en strategieën, met name door middel van het opgebouwde uitvoeringskennis en netwerken. Hieruit kan worden geconcludeerd dat governance voor de strategische planning en governance voor de realisatie van adaptatiemaatregelen elkaar kunnen versterken. De wetenschap over klimaatadaptatie heeft dus baat bij een meer uitvoeringsgerichte benadering van adaptatie.

Table of contents

CHAPTER ONE

Introduction

Overcoming hurdles to adaptation.

This chapter describes that adaptation action to flooding and drought is being impeded by governance challenges. Guidance for governance of adaptation is, therefore, needed to overcome challenges to adaptation action. This thesis aims to address this problem by developing prescriptions for different aspects of governance of adaptation, whilst ensuring that governance approaches are appropriate and specific within their context. Therefore, the overarching research question of this thesis is:

How can adaptation actions to manage changes in flood and drought risks be delivered effectively?

This chapter describes that this thesis addresses the overarching question through focusing on the questions of what is effective governance of adaptation and when, why and how governance can enhance the uptake of adaptation action. Furthermore, the main contributions and structure of this thesis are described in this chapter.

1. Introduction

1.1 Background

There is a great need for societies to adapt to climate change in order to anticipate increasing risks and/or seize new opportunities (e.g. EEA, 2012; European Commission, 2013; UNISDR, 2013; World Bank, 2013). For example, the European Commission states the following in its recent *EU strategy on adapting to climate change* (European Commission, 2013; p. 3-4):

> *"The minimum cost of not adapting to climate change is estimated to range from €100 billion a year in 2020 to €250 billion in 2050 for the EU as a whole (EEA, 2012). Between 1980 and 2011, direct economic losses in the EU due to flooding amounted to more than €90 billion (EEA, 2012). This amount is expected to increase, as the annual cost of damage from river floods is estimated at €20 billion by the 2020s and €46 billion by the 2050s (Rojas et al., 2013). The social cost of climate change can also be significant. Floods in the EU resulted in more than 2500 fatalities and affected more than 5.5 million people over the period 1980-2011. Taking no further adaptation measures could mean an additional 26 000 deaths/year from heat by the 2020s, rising to 89 000 deaths/year by the 2050s (Kovats et al., 2011). Though there is no real comprehensive overview of adaptation costs in the EU, additional flood protection measures are estimated at €1.7 billion a year by the 2020s and €3.4 billion a year by the 2050s (Feyen and Watkiss, 2011). Such measures can be very effective, as for each euro spent on flood protection, we could avoid six euros of damage costs (Feyen and Watkiss, 2011)."*

This quotation also illustrates that the climate impacts and adaptation costs are uncertain. In fact, it is acknowledged that it is impossible to precisely predict the future climate and its impact upon society (Cox and Stephenson, 2007; Milly et al., 2008). In many places, the acknowledgement of increasing risks and their related uncertainties are forcing fundamental reforms of the way societies manage their water systems in order to establish and/or maintain adequate provision of flood protection, but also other functionalities such as water security and environmental protection (e.g. Deltacommissie, 2008; Evans et al., 2004; NWC, 2007). Research has demonstrated that such reforms should result in adaptive and resilient forms of water management that explicitly take into account the uncertainties of climate induced risks on the immediate and long term time scales (e.g. de Bruijn, 2005; Folke et al., 2005; Rockström, 2003). Resilient water systems can be classified as systems that have the capacity to absorb shocks whilst maintaining function, and to recover and re-organise after a shock has taken place (Folke, 2006; Gersonius et al., 2010).

The IPCC defines adaptation as *"the adjustment in natural or human systems in response to actual or expected climatic stimuli or their effects, which moderates harm or exploits beneficial opportunities"* (Parry et al., 2007, p.6). Adaptation to climate change is commonly referred to as a governance challenge (e.g. Adger et al., 2009; Folke, 2006; OECD, 2011). With regard to water management, the focus of this thesis, this governance challenge can be broken down into several parts. Firstly, the practical implementation of available innovative technologies and knowledge required to develop adaptive water management systems is slow (Harding, 2006; Mitchell, 2006). Secondly, it is nowadays frequently suggested that new modes of governance are needed that are effective under conditions of high complexity and uncertainty (e.g. Folke et al., 2005; OECD, 2011; Zevenbergen et al., 2012). These approaches would involve, for example, multiple disciplines, multiple government levels, the community, the private sector and academia. Adger and colleagues summarise these governance challenges nicely, by suggesting that *"adaptation to climate change is limited by the values, perceptions, processes and power structures within society"* (Adger et al., 2009, p.349).

Successful adaptation is not a one-off activity: adaptation measures should be considered as temporary responses rather than definitive solutions, because it is likely that their functionality changes over time as a result of, for example, climate change, economic growth and demographic change (Adger et al., 2005b). This implies that governance of adaptation has a twofold objective of: 1) overcoming the impediments to taking adaptation measures; and 2) enhancing society's capacity to remain adaptive to change after adaptation measures have been taken. Both objectives are implicitly captured in the title of this thesis, 'Delivering change', as a new approach to governance is needed to deliver ongoing adaptation effectively.

1.2 Problem and scope

Summarising the above, it could be concluded that adaptation action is being impeded by governance challenges. Guidance for governance of adaptation is, therefore, needed to overcome challenges to adaptation action. This thesis addresses this demand by focusing on the governance throughout the adaptation process from intentions for adaptation to the execution and realisation of adaptation action. In this section, the scope of the thesis is described through a brief overview of some key concepts and the challenges that they currently provide to the research about the governance of adaptation. More detailed descriptions of the key concepts are given in the chapters 2 to 6.

1.2.1 Transformational processes: adaptation and transitions

In this thesis, I use the terminology of transformational processes, or transformation, to refer to processes in which the form or functionality of systems change. With the above definition of the IPCC for adaptation (section 1.1), adaptation can be considered as an example of a transformational process. Another example of a transformational process is a transition, which is a structural change in the way a society or a subsystem of society (e.g. water management, energy supply, agriculture) operates, and which can be described as a long-term non-linear process (25-50 years) that results from a co-evolution of cultural, institutional, economic, ecological and technological processes and developments on various scale levels (Rotmans et al., 2001). Whilst adaptation is not necessarily a permanent change, transitions are by definition structural changes of practices, institutions and culture. However, this thesis focuses on adaptation of water systems through technological interventions in the existing water infrastructure systems. In the context of this thesis, the process of adaptation is practically similar to transition, because these interventions alter the existing infrastructure systems structurally. In addition, adaptation and transitions are both the result of self-organisation and/or deliberate planning and, in case of the latter, they both require continuous influence and adjustment in governance systems (Foxon et al., 2009; Smith and Stirling, 2010). For this reason, I use the terminology of transformation, adaptation and transition interchangeably.

1.2.2 Governance

Governance is a concept rooted in the social sciences and as such is defined and interpreted in many different ways (for an overview of definitions and interpretations, see e.g. Kjær, 2004; Rhodes, 1996). Governance incorporates both processes and structures required for steering and managing parts of societies (Kooiman, 1993; Pierre and Peters, 2000). As a process, governance refers to managing networks, markets, hierarchies or communities (Kjær, 2004; Rhodes, 1996), whereas governance as structure refers to the institutional design of patterns and mechanisms in which social order is generated and reproduced (Voß, 2007). Taking a combined view, governance can be considered as comprising three mutually reinforcing elements: policy (problems and solutions), polity (rules and structures), and politics (interaction and process) (Voß and Bornemann, 2011). Governance is also the outcome of interaction among multiple actors from different sectors with different levels of authority (multi-level governance; Agrawal, 2003). As such, governance relies on institutions consisting of cognitive (dominant knowledge, thinking and skills), normative (culture, values and leadership) and regulative components (administration, rules and systems) that mutually influence practice (Scott, 2001).

It is a common critique amongst researchers focusing on governance of adaptation that there is insufficient prescription for transformative governance approaches that are able to assist practitioners to enhance resilient water systems effectively (e.g. Huitema et al., 2009; Loorbach, 2010). Creating effective prescription is complicated by the recognition that there are no blueprint solutions for good governance that operate successfully in all conditions and across all scales (Ostrom et al., 2007; Pahl-Wostl et al., 2010). However, several recent contributions have been useful for developing prescriptions for effective governance through guiding principles (e.g. Huntjens et al., 2012; Ostrom and Cox, 2010) and attributes of transformative governance (e.g. Farrelly et al., 2012; Loorbach, 2010; Pahl-Wostl et al., 2010; van de Meene et al., 2011). Whilst all these efforts provide general guidance for policy and decision makers to governance arrangements that enhance resilience, most of them fail to provide specific guidance for governance related to changing circumstances during transformation processes, with some recent exceptions (i.e. Adger et al., 2011; Herrfahrdt-Pähle and Pahl-Wostl, 2012; Olsson et al., 2006). This issue is taken as a point of departure for this thesis to enrich the scientific knowledge related to governance of adaptation and develop practical guidance for the governance of adaptation.

1.2.3 Strategic planning for adaptation and delivery of adaptation action

Within the setting of complex systems, adaptation can be induced by self-organisation or deliberate planning (section 1.2.1). This thesis considers the entire process of deliberately planned adaptation; from intentions for adaptation to the delivery of adaptation action. With regard to deliberately planned adaptation, adaptation can be considered as a continuous cycle of activities for understanding the need for adaptation, planning for adaptation and managing adaptation action (Moser and Ekstrom, 2010). Because implementation of adaptation in practice is the primary concern of this thesis, I distinguish between strategic planning for adaptation and delivery of adaptation. In this thesis, strategic planning for adaptation refers to activities that relate to understanding the need for adaptation and planning for adaptation action. Delivery of adaptation refers in this paper to managing adaptation action in practice after a particular action has been selected. Recent studies in, for example, Australia, the UK and Scandinavia, have shown that adaptation research findings are often not being adopted in practice (Brown et al., 2011; Klein and Juhola, 2013). It is suggested that slow uptake of adaptation research into practice can be attributed to a dominant research focus on system performance which is currently neglecting the perspective of decision-makers and the role of agency (Klein and Juhola, 2013). Moreover, recent research has pointed out that the considerable research efforts on adaptation mainly focuses on intentions to adapt rather than on real adaptation actions in practice (Berrang-Ford et al., 2011). This suggests that governance for the delivery of adaptation action is largely overlooked by adaptation research. Filling this void, this thesis considers governance for strategic planning of adaptation, governance for delivery of adaptation and the interaction between both elements.

1.3 Aim and research questions

This thesis aims to address this problem, developing prescriptions for different aspects of governance of adaptation, whilst ensuring that governance approaches are appropriate and specific within their context. It specifically focuses on adaptation to flooding and drought. Therefore, the overarching research question of this thesis is:

How can adaptation actions to manage changes in flood and drought risks be delivered effectively?

This thesis addresses the overarching question through focusing on the questions of what is effective governance of adaptation and when, why and how governance can induce and enhance the uptake of adaptation action:

- WHAT: What is effective governance of adaptation?
- WHEN: When, during different stages of transformation, is a particular governance approach effective?
- WHY: Why are transformational processes sometimes being hampered?
- HOW: How can deliberate adaptation projects be realised effectively?
- HOW: How can strategic planning enhance the implementation of adaptation action effectively?

These research questions are addressed separately in chapters 2 - 6 of this thesis (see section 1.5). Hence, for each of the questions, the introduction, background theory, methodology, research findings, discussion and conclusions are described separately. As this is a thesis-by-papers, each of these chapters can be read independently without prior knowledge of the others.

1.4 Research approach

My motivation to write this PhD thesis is mainly driven by the desire to assist in addressing 'real' problems of practitioners and policy makers. In my view, science has an important role to inform practice about emerging issues and alternative practices. With respect to governance, I perceive the role of science as a means to explain the outcomes of past and current governance approaches and structure policy debates about future governance approaches.

I hold a MSc degree in civil engineering with a specialisation in water resources management (TU Delft). During my BSc and MSc studies, I was trained to develop technological solutions to problems that were often preset with clear boundaries and limited uncertainties. Assignments included, for example, the design of a sewage pipe or irrigation channel from point A to B for a given probability of design discharge capacities. However, during exchange programmes within my MSc at respectively KTH in Stockholm and Monash University in Melbourne, I was introduced to social sciences and learned about the design of processes to deliver technological systems and the importance of understanding the context for developing appropriate technological systems.

In this thesis, I use the solution oriented focus of my engineering background within the social sciences domain. As described above, this thesis provides guidance for delivering adaptation action effectively. Although the technological concepts in this thesis are often considered 'best practice' or 'sustainable', this thesis does not attempt to provide evidence that supports their effectiveness nor does it argue that these technological concepts are better than others. The value judgement about the effectiveness and appropriateness of the technological concepts discussed should be considered by engineers and social scientists respectively. Instead, this thesis focuses on how governance can stimulate the uptake of these technological concepts.

This thesis reflects the outcomes of demand-driven research: practical questions were the incentive for starting the research projects that provided the input for this thesis. The conducted research was therefore primarily steered by practice and subsequently shaped by theory. I have combined different theories, for example about transition management, adaptive governance and project management, to structure and interpret empirical findings and develop new insights about the governance of adaptation.

As a logical consequence of the demand-driven research approach, I actively engaged with the key stakeholder groups during the entire research process. Stakeholders were involved with identifying the research aim, they participated as respondents in the data collection, and I discussed my preliminary findings with them during the validation phase of the research. As such, my research influenced the uptake of the technologies by informing these stakeholders and engaging them in a dialogue about the uptake of innovative technological concepts. However, it cannot be determined how influential the research projects of this thesis exactly were on the perceptions and actions of stakeholders with whom I have engaged.

My active engagement with the key stakeholder groups has had several implications for this thesis. It has directed my research towards relevant research questions and applicable outcomes. However, caution needed to be taken to avoid research outcomes that were biased towards stakeholders' preconceptions and/or interests. In the research project that focused upon urban water governance in Australia (chapters 3 and 6), objectivity was safeguarded by involving a broad array of stakeholders during all stages of the research. The same strategy was used in the scientific evaluation of the Room for the River programme in the Netherlands (chapters 4-6). In addition, the scientific rigour of this research was evaluated by a scientific advisory board whilst the practical relevance of the research was reflected upon by a user panel consisting of experts with policy and advisory roles.

1.5 Overall research design

Figure 1.1 summarises the overall design of the research that is presented in this thesis.

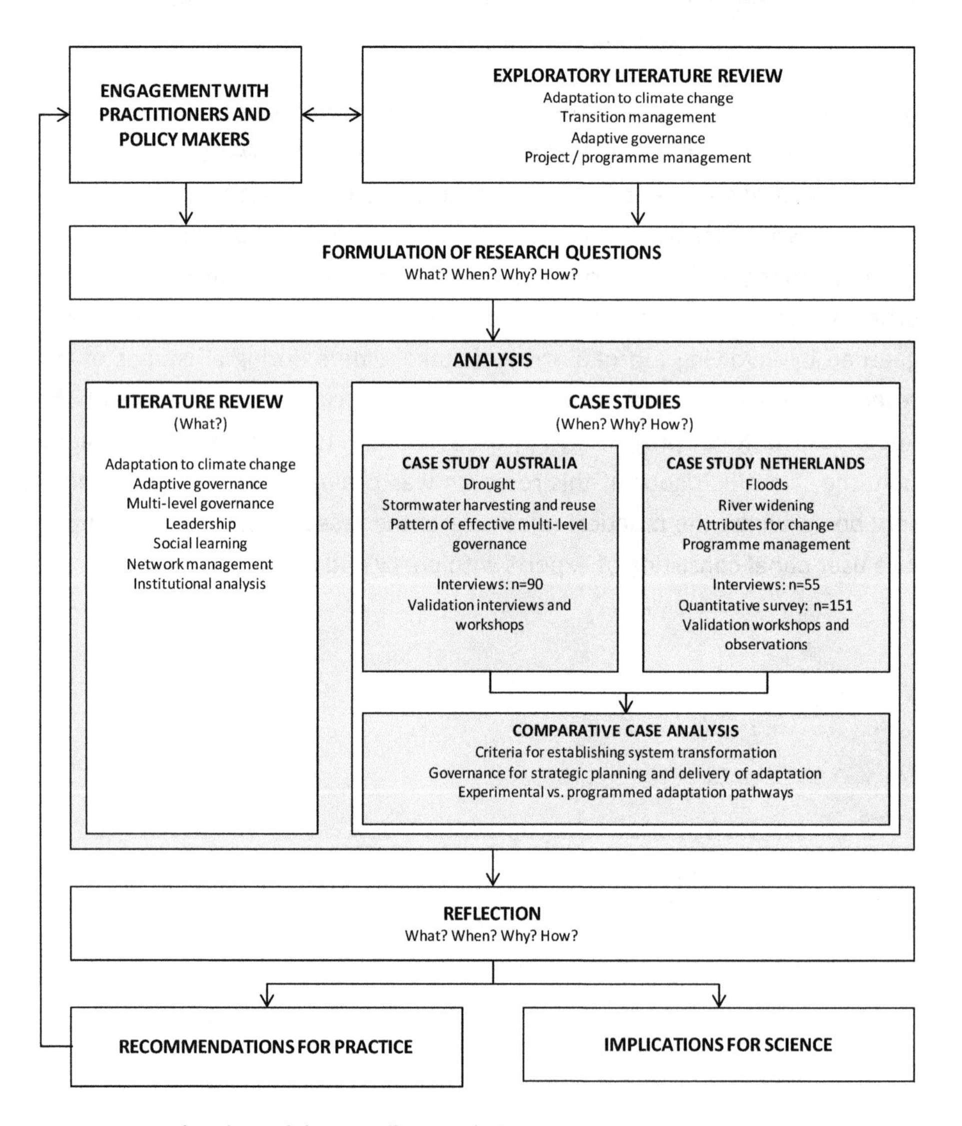

Figure 1.1 Flowchart of the overall research design

The research questions (section 1.2) are formulated after a combination of active engagement with stakeholders and exploratory literature reviews. As explained in section 1.3, this thesis presents the outcomes of demand-driven research that was requested by practice and shaped by theory. As a consequence, the research questions are formulated to answer practical and theoretical needs. Initial discussions with the stakeholder groups who commissioned the two research projects in Australia and the Netherlands are held, particularly on the outset of the research, to match the scope of the research with the stakeholder needs. In addition, literature reviews are conducted to explore the theoretical background of the issues raised by the stakeholders, determine the scope of the two respective research projects and safeguard their scientific relevance by matching the practical issues with scientific challenges which are described in the literature. In line with the aim of this thesis, particularly the bodies of literature related to adaptation to climate change (in the water sector), transition management, adaptive governance and project/programme management are drawn upon to frame the two separate research projects that informed this thesis. The applied approach enabled to formulate the research questions in a way that they are relevant to both science and practice.

The analysis that is conducted to answer the research questions consists of a literature review and two separate case studies. The selected case studies include the analysis of governance for the uptake of stormwater harvesting and reuse as a means to adapt to drought in Australian cities and the analysis of governance for the delivery of river widening as a means to adapt to flooding in the Netherlands. These cases are selected because they are on a global scale considered relatively successful in terms of the delivery of adaptation through innovative water management approaches (for the Australian case, see e.g. Farrelly and Brown, 2011; Howe and Mitchell, 2011; for the Dutch case see e.g. van Herk et al., under review; Warner et al., 2013). As such, they provide the opportunity to identify lessons for turning intentions for adaptation into concrete adaptation actions through strategic planning and realisation processes. I have deliberately avoided case studies with failing governance approaches for adaptation, because numerous studies of barriers and challenges to adaptation, particularly in the water sector, have been published during the last few years (e.g. Adger et al., 2009; Biesbroek et al., 2013; Brown et al., 2009a; Jones and Boyd, 2011). Furthermore, both cases are complementary, as they cover adaptation to drought (Australia) and flooding (Netherlands) and because the processes of adaptation in the two cases have different pathways (experimental vs. programmed; see chapter 6). As a result, the combination of the two cases provides the opportunity to compare criteria for establishing system transformation and effective governance approaches for strategic planning and delivery of adaptation for different adaptation pathways.

More specifically, the combination of the case studies and extensive literature review is being used to answer the research questions that are described in section 1.3. The question of what is effective governance of adaptation is primarily being addressed through a review of the literature related to adaptation to climate change, adaptive governance, multi-level governance, leadership, social learning, network management and institutional analysis (chapter 2).

The question of when during different stages of transformation a particular governance approach is effective, is being addressed through a comparison of the transformation processes in three cities in the Australian urban water management context (chapter 3). The distinction of these three separate (sub-)cases within the Australian stormwater management case provides the opportunity to investigate the effectiveness of governance approaches in different stages of transformation, but with an otherwise mostly similar institutional, technological and climatic context.

The question of why transformational processes are sometimes being hampered is being answered through validation of a framework of attributes for transition governance that is developed and applied in the Australian urban water context (see Appendices A and B). Application of the existing framework in the Dutch context of river widening (chapter 4) is used to confirm/reject/nuance the existing framework and thus increase the understanding of underlying reasons for barriers to transformation.

Furthermore, the Dutch case provides the opportunity to analyse how deliberate adaptation action can be realised effectively, because river widening measures are being implemented on a system scale in a dedicated investment programme without (thus far) overrunning its dedicated budget of 2.4 billion Euro (i.e. the Room for the River programme). The case of the Room for the River programme is being used to verify and enrich an existing framework for effective programme management (chapter 5).

Finally, the two cases of governance of adaptation in Australia and the Netherlands are being compared to address the research question of how strategic planning can enhance the implementation of adaptation action effectively (chapter 6). The relative successful implementation of adaptation measures and the different characteristics of the adaptation pathways in both cases (experimental vs. programmed; see above) provides the opportunity to analyse how governance of strategic planning for adaptation can enhance the governance of the delivery of adaptation, and vice versa.

For each of the sub-analyses that address the research questions, I reflected on the possible application and the limitations of the research findings to identify the key recommendations for practice and the implications for further research (chapter 7).

1.6 Contributions

This thesis contributes to practice and science. The main contributions are outlined below and reflected upon in chapter 7. The main contributions of this thesis are:

- A procedure is developed for assessment of the 'fit-for-purpose' of applied and proposed governance approaches for enhancing adaptation action. By assisting policy and decision makers in avoiding panaceas and customising institutional reform, this procedure provides the basis for a new way of thinking to address impediments to the uptake of adaptive governance. See chapter 2.
- Through application of the 'fit-for-purpose' governance procedure, a pattern of effective governance configurations during consecutive stages of transformation processes is identified. This pattern can be used to provide guidance for urban water governance reform to policy makers and governance evaluators. See chapter 3.
- Through application of an existing 'transitions governance' framework, a set of criteria for establishing structural system change (i.e. a transition) is tested for the context of river flood protection in the Netherlands. These criteria can be used as a checklist for policy and decision makers to establish systemic transformations, such as transitions and system-wide adaptation. See chapter 4.

- A set of attributes for the effective delivery of large scale adaptation projects. These attributes are specifically developed for policy makers, programme managers and project managers who are involved in setting up and managing large flood protection programmes. See chapter 5.
- Insight into the interaction between governance for strategic planning and delivery of adaptation. Based on a comparative study of adaptation in the water sectors in the Netherlands and Australia, I conclude that the uptake of planned adaptation action can be stimulated through reinforcing connections between the governance for strategic planning of adaptation and the governance of delivery of adaptation projects. See chapter 6.

1.7 Thesis structure

Figure 1.2 shows how this thesis is structured around the main research questions and contributions that are outlined above. Each of the chapters 2-6 consists of a scientific journal paper that has recently been published (three published papers) or is currently under revision (two papers under revision). The synthesis (chapter 7) summarises the key findings of each of the papers, reflects on the collection as a whole and draws conclusions and recommendations therefrom.

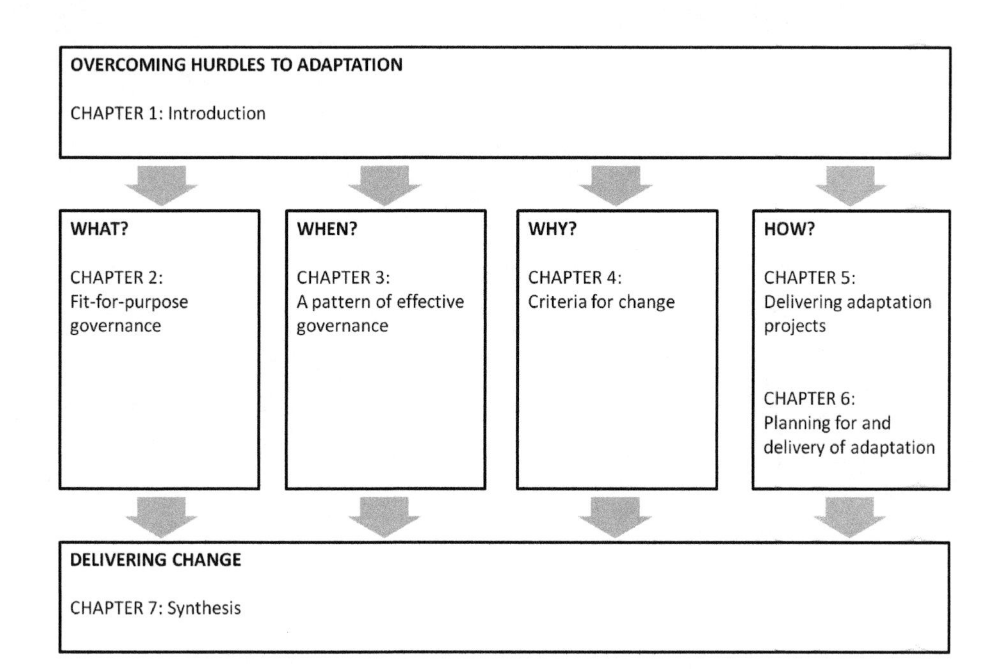

Figure 1.2 Thesis structure

CHAPTER TWO

Fit-for-purpose governance: a framework to make adaptive governance operational

Fit-for-purpose governance.

This chapter provides a procedure that assists policy makers and practitioners in assessing how 'fit-for-purpose' an applied or proposed governance approach is under a given set of contextual conditions. By assisting policy and decision makers in avoiding panaceas and customising institutional reform, this procedure provides the basis for a new way of thinking to address impediments to the uptake of adaptive governance. As such, it relates to the 'what' question: *What is effective governance of adaptation?*

2. Fit-for-purpose governance: a framework to make adaptive governance operational

This chapter is adapted from:

Rijke. J., Brown, R., Zevenbergen, C., Ashley, R., Farrelly, M., Morison, P. and van Herk, S. (2012) Fit-for-purpose governance: a framework to make adaptive governance operational. *Environmental Science & Policy*, 22: 73-84.

Abstract

Natural disasters, extreme weather events, economic crises, political change and long term change, such as climate change and demographic change, are in many places forcing a re-think about the way governments manage their environmental resource systems. Over the last decade, the concept of adaptive governance has rapidly gained prominence in the scientific community as a new alternative to the traditional predict-and-control regime. However, many policy makers and practitioners are struggling to apply adaptive governance in practice. Drawing on an extensive, critical literature review of adaptive governance, network management and institutional analysis, I argue that the constraints to the uptake of adaptive governance relate to a large extent to the inability of practitioners and policy makers to cope with complexity and various uncertainties: i) ambiguous purposes and objectives of what should be achieved with governance; ii) unclear contextual conditions in which governance takes place; and, iii) uncertainty around the effectiveness of different governance strategies. To address such practical challenges, this chapter introduces a 'fit-for-purpose' framework consisting of three key ingredients for developing a diagnostic approach for making adaptive governance operational. I introduce the concept of fit-for-purpose governance to be used as an indication of the effectiveness of governance structures and processes and define it as a measure of the adequacy of the functional purposes that governance structures and processes have to fulfil at a certain point in time. In other words, are existing and proposed governance structures and processes fit for their purpose? While adaptive governance focuses on responding to (potential) change, fit-for-purpose governance is specifically considering the (future) functions that the social and physical components of a particular system, such as an urban water system, have to fulfil. As such, the fit-for-purpose governance framework provides an alternative starting point for developing the much sought-after guidance for policy and decision makers to evaluate the effectiveness of established governance arrangements and to predict the likelihood of success of institutional reform.

2.1 Impediments to the implementation of adaptive governance

Natural disasters, extreme weather events, economic crises, political change and long term change, such as climate change and demographic change, are in many places forcing a re-think about the way governments manage their environmental resource management systems. For example, adaptation to climate change is commonly referred to as a governance issue (e.g. Adger et al., 2009; Adger et al., 2005a; Folke, 2006). Developing resilient governance systems to manage environmental assets to support secure, long-term societal development is challenging (Costanza et al., 2000; Lambin, 2005). Research has demonstrated that this challenge requires adaptive forms of governance that explicitly take in to account immediate and long term change (Dietz et al., 2003; Folke et al., 2005). However, the complexity of system dynamics and interactions between different components of governance systems causes inherent uncertainty in terms of short, medium and long term outcomes. Therefore, adaptive governance attempts to address uncertainty through continuous learning, involvement of multiple actors in decision making processes and self-organisation of the governance system.

Continuous learning is a critical component of adaptive governance in order to be able to take into account complex dynamics and uncertainty (e.g. Folke et al., 2005). Learning processes are stimulated by networks that enable interaction between individuals, organisations, agencies and institutions at multiple organisational levels to draw upon various knowledge systems and the experience to develop policies (e.g. Adger, 2001; Adger et al., 2005a; Olsson et al., 2006). Adaptive governance relies on polycentric institutional arrangements that operate at multiple scales (McGinnis, 1999; Ostrom,

1996), and balance between centralised and decentralised control (Imperial, 1999). Furthermore, adaptive governance systems often self-organise as a result of learning and interaction (e.g. Folke, 2003). However, self-organisation needs to be enabled by flexible institutional arrangements that encourage reflection, innovative responses, and some redundancy (Brunner et al., 2005; Folke et al., 2005; Pahl-Wostl, 2006). Leadership of individuals or organisations may serve as a catalyst for emergent adaptive processes by strategically bringing together people, resources and knowledge (e.g. Boal and Schultz, 2007; Lichtenstein and Plowman, 2009; Uhl-Bien et al., 2007).

The technologies and knowledge required to develop adaptive environmental resource management systems are in most cases available, but their implementation into practical action remains slow (Harding, 2006; Mitchell, 2006). Numerous scholars have identified a range of impediments, many of them related to governance (e.g. Brown and Farrelly, 2009a; Maksimovic and Tejada-Guilbert, 2001). For example, Australian urban water practitioners who have tacit knowledge of the operation of traditional systems are insufficiently engaged in policy making to incorporate practical knowledge about opportunities and impediments for more sustainable water management (Brown et al., 2009a). Furthermore, recent research demonstrates practitioners are willing to embrace new practices but are currently constrained by, among other things, traditional servicing arrangements, limited capacity (skills and knowledge of new technologies / systems / practices) and concerns regarding the potential risks to public health and welfare (Brown et al., 2009a; Farrelly and Brown, 2011).

This chapter aims to assist in overcoming the challenges of making adaptive governance operational by providing a tentative framework for policy practitioners and decision makers for assessing the effectiveness of governance approaches. This 'fit-for-purpose' governance framework provides the ingredients for assessing the effectiveness of existing and proposed governance mechanisms to fulfil their purpose in a particular context. The framework was developed after an in-depth review of the underlying reasons that cause challenges in practice in the institutional science and (adaptive) governance literatures related to environmental resource management (Section 2.2). This revealed that constraints to the uptake of adaptive governance relate, to a large extent, to the inability of practitioners and policy makers to cope with complexity and uncertainties. Several efforts have been made to develop principles for effective governance of social-ecological systems (e.g. Huntjens et al., 2012; Ostrom and Cox, 2010). However, in practice a tendency to implement panaceas for the governance of social-ecological systems has been observed in the past (Ostrom et al., 2007). Using the literature on policy analysis related to social-ecological systems, the fit-for-purpose framework is developed as a diagnostic procedure that can guide policy practitioners through a logical process, while the framework itself reflecting contemporary and adaptive understandings of governance. Drawing upon literature bodies related to networks, leadership and social learning, a first attempt is made to make the fit-for-purpose framework operational (Section 4). Furthermore, the potential applications and limitations of the fit-for-purpose governance framework are discussed (Section 2.5).

2.2 Three uncertain aspects that create challenges for adaptive governance

Drawing on insights gained from an extensive, critical literature review on adaptive governance, network management and institutional analysis, I argue that constraints to the uptake of adaptive governance relate, to a large extent, to the inability of practitioners and policy makers to cope with complexity and uncertainties. In particular: i) ambiguous purposes and objectives of what should be achieved with governance; ii) unclear contextual conditions in which governance takes place; and iii) uncertainty around the effectiveness of different governance strategies.

2.2.1 Ambiguous purposes of governance

According to many scholars, there is a shift taking place from government to governance; a shift from hierarchical and well-institutionalised forms of governance performed by a dominant bureaucratic and administrative government, to less formalised governance approaches with power distributed amongst various actors and organisations (e.g. Arts et al., 2006; Hanf and Scharpf, 1978; Ostrom, 1990). Governance is a concept that is defined and interpreted in many different ways (for an overview of definitions and interpretations, see e.g. Kjær, 2004; Rhodes, 1996). It refers to both processes and structures for steering and managing parts of societies (Kooiman,

1993; Pierre and Peters, 2000; see also van Nieuwaal et al., 2009). Governance as process refers to managing networks, markets, hierarchies or communities (Kjær, 2004; Rhodes, 1996). In this sense, governance refers to governing and can be defined as "the setting, application, and enforcement of the rules of the game" (Kjær, 2004, p. 12), or as "all those activities of social, political and administrative actors that can be seen as purposeful efforts to guide, steer, control or manage (sectors or facets of) societies" (Kooiman, 1993, p. 2). Governance as structure refers to the pattern of institutional design and the mechanisms in which social order is generated and reproduced (Voß, 2007). In this respect, governance is defined as "the patterns that emerge from governing activities of social, political and administrative actors" (Kooiman, 1993, p. 2). Here, I take into account both interpretations of governance and consider it as the total of: the networks of actors, institutional frameworks and processes that take place within these networks and frameworks.

Identifying the purpose of governance is not straightforward (see also Adger et al., 2009; Smith et al., 2005). For example, the official objective of the 2.3 billion Euro flood protection program Room for the River in the Netherlands was set by the Dutch Government in December 2006 to increase the discharge capacity of the river systems to 16.000 m^3/s by 2015, whilst contributing to spatial quality of the river landscape (www.roomfortheriver.nl). The ambiguity arises from the second part of the objective, because different stakeholders may have different ideas about 'contributing to spatial quality'. For example, certain stakeholders may prefer new opportunities for development, whilst others pursue the creation of nature and/or recreation areas. According to Adger et al (2009, p. 339), such diversity of values may often lead to "a paralysis of adaptation actions". Furthermore, the ambiguity of governance purposes raises questions such as "who governs?" and "whose sustainability gets prioritised?" (Smith and Stirling, 2010). Hence, it can be concluded that ambiguous governance purposes resulting from a range of values creates a significant challenge for applying adaptive governance.

2.2.2 Unclear governance context

Social-ecological systems can be described as complex adaptive systems that evolve through interaction between social and natural sub-systems (see also

Berkes et al., 2000; Folke, 2006). Interactions between the physical components of the social-ecological system, the governance system and the users of, for example, the urban water system, result in outcomes that evolve in time and space (Ostrom, 2007). Hence, changing conditions in the social and physical context of social-ecological systems influences the effectiveness of governance to serve a specific purpose. Governance, and adaptive governance in particular, relies on networks that connect actors (individuals, organisations, agencies, and/or institutions) at multiple organisational levels (Folke et al., 2005). The effectiveness of networks to solve complex problems, such as adaptive governance of environmental resource systems, depends on the combination of network structure and context (Turrini et al., 2010). Research undertaken in the computer sciences has shown that the concept of context is generally understood by a set of circumstances that frame an event or object, but it remains ill-defined in the cognitive and related sciences (Bazire and Brézillon, 2005). Several frameworks from the literatures about institutional analysis (e.g. Kiser and Ostrom, 1982; McGinnis, 2011), transition management (e.g. Geels, 2002; Rotmans et al., 2001) and adaptive governance (e.g. Pahl-Wostl, 2007) provide key components for mapping the context, such as rules, dominant paradigms, available technology and knowledge and biophysical conditions. However, as Ostrom (2011) comments, a framework merely identifies elements and general relationships that need to be considered for institutional analysis. It does not provide analysts nor practitioners specific methods for how a context can be mapped in order for it to establish effective governance strategies. This reveals the need for further work to operationalise adaptive governance in the future in order to be able to better predict the likelihood of success of adaptation measures.

2.2.3 Uncertain governance outcomes

As mentioned above, governance relies on networks that connect actors at multiple organisational levels. Thus, analysing relations between actors helps to understand how social structures (the regime) enhance or hinder effective governance. Turrini et al (2010) suggest that the effectiveness of networks to solve 'wicked' problems such as adaptation to climate change depends on a combination of network structure and context. However, Ostrom et al.

(2007) argue that practitioners and scholars have a tendency towards developing panacea, blueprint solutions, to all types of environmental problems and fail to take uncertainty and the complex dynamics of governance systems into account. For example, the privatisation of public services or decentralised management of natural resources have a track record of repeated failure related to unanticipated outcomes (Acheson, 2006). Therefore, it is not surprising that in many developed countries a paradigm shift is currently taking place in water governance from "a prediction and control to a management as learning approach" (Pahl-Wostl, 2007, p.49). Prediction and control approaches are derived from mechanistic thinking in which system behaviour and response can be predicted and optimal control strategies can be designed within regulatory frameworks that are shaped by technical norms and legal prescriptions (Pahl-Wostl, 2007). Management as learning approaches are essentially adaptive approaches derived from complexity and resilience thinking in which self-organisation and learning have a central place (Pahl-Wostl, 2007). Such learning approaches embrace uncertainty by iterative processes of adjusting governance to achieve better outcomes over time. However, policy makers and practitioners continue to struggle with setting learning goals and expectations, defining adequate learning mechanisms, and identifying who should be involved in learning processes (Armitage et al., 2008). This hampers their ability to develop adaptive governance strategies which rely on continuous learning.

2.3 Proposal for a framework to overcome challenges for adaptive governance

Adaptive governance offers an important theoretical framework for developing more sustainable governance of environmental resources, but needs to be supported by tools for operationalisation. In engineering, examples of supporting tools to help put adaptation in practice such as the 'adaptation tipping point' method (Kwadijk et al., 2010) or 'real options' analysis (Gersonius et al., 2010) are readily available. However, supporting tools are still required to shift adaptive governance from rhetoric to practice. Water management and climate adaptation practice and policy making in, for example, Australia (Nelson et al., 2008) and the Netherlands (Anema and Rijke, 2011) are facing difficulties in putting the principles of adaptive governance into practice. In particular embracing complexity and uncertainty, continuous learning, and ongoing reflection and adjustment of management approaches, are providing practical challenges because they are not being institutionalised into planning practice. According to practitioners and policy makers, adaptive approaches should preferably be incorporated into existing institutional frameworks in order to achieve such a shift (Rijke et al., 2009). However, most existing institutional frameworks are based on the predict and control paradigm and act as the institutional expression of reducing uncertainty (see also Pahl-Wostl, 2007). As such, they are designed to provide 'optimal' solutions to environmental resource problems. Inherent uncertainty of climate behaviour (Milly et al., 2008), alongside the uncertainties of adaptive governance that are described above, make development of such solutions practically impossible. Hence, there is a mismatch between the existing institutional frameworks in which policy makers and practitioners operate and the principles of adaptive governance, such as flexibility and self-organisation (see also Nelson et al., 2007).

To address the challenges to operationalise adaptive governance, I propose a complementary framework that uses dominant institutional arrangements rather than flexibility and self-organisation as the starting point. However, rather than aiming for good or even 'optimal' governance, it aims for "good enough governance", which takes into account uncertainty by focusing on essential adjustments, priorities in the short and long term and feasibility, and therefore may be a more realistic goal (Grindle, 2004; p.526). In order to operationalise the concept of adaptive governance and avoid the pitfalls of panacea (Section 2.3), I propose a framework that provides the ingredients for assessing the effectiveness of existing and proposed governance mechanisms to fulfil their purpose in a particular context. In other words, are existing and proposed governance mechanisms (governance structures and processes) fit for their purpose? Applying such a diagnostic approach provides insight about how particular solutions improve or aggravate outcomes and assists in avoiding developing inadequate governance solutions (Ostrom, 2007; Pahl-Wostl et al., 2010). Assessment of the impact of particular solutions requires knowledge about the purpose for which these solutions are implemented and the context in which they are implemented. As such, the fit-for-purpose framework provides guidance for establishing fit-for-purpose governance.

I define fit-for-purpose governance as a measure of the adequacy of the functional purposes that governance structures and processes have to fulfil at a certain point in time. A fit-for-purpose governance structure (e.g. a hierarchy or a free market) enables social, political and administrative actors to purposefully guide, steer, control or manage (sectors or facets of) societies through network structures that have a fit to their physical and social context (adapted from Kooiman, 1993). Fit-for-purpose governance processes (e.g. leadership or social learning) are fit to both the network structure in which they take place and the purpose for which they are being used. While adaptive governance focuses on responding to (potential) change, fit-for-purpose governance is specifically considering the (future) functions that the social and physical components of a particular social-ecological system have to fulfil. In other words, adaptive governance is about ongoing action while fit-for-purpose governance is an indication of the effectiveness of such action. Therefore, the two concepts are complementary and using them concurrently creates synergies: the concept of fit-for-purpose governance may provide the much sought-after guidance for policy makers and decision makers to predict the likelihood of success of institutional reform by diagnosing the fit of governance arrangements with the purpose for which they are being proposed or applied. Subsequently, learning processes characteristic to adaptive governance could use the results of such diagnosis to evaluate the effectiveness of governance in relation to any immediate crises and/or long-term change.

In Figure 2.1, a three-step framework to diagnose the fit-for-purpose of governance mechanisms is presented. By making the three uncertain aspects that create challenges for the operationalisation of adaptive governance explicit, the framework aims to make policy makers and decision makers aware of issues that need to be resolved in order to develop effective (adaptive) governance mechanisms. As such, the fit-for-purpose framework identifies ingredients from which a tool for establishing adaptive governance can be developed. First, the purpose of implemented or proposed governance mechanisms needs to be identified in terms of policy objectives (e.g. expressed by temporal and spatial dimensions and/or production, consumption flow of resources). Secondly, the context in which governance strategies are implemented needs to be mapped. Despite the lack of available tools to map a particular context, frameworks are developed that provide a starting point for doing so (see Section 2.2). For example, a governance system can be considered as a subsystem of a social-ecological system that interacts with: (1) resource systems (e.g. sewage systems, rivers) in which resource units (e.g. wastewater, fish) are produced, consumed or transported; (2) related ecosystems; and (3) social, economic and political settings (Ostrom, 2007). Hence, it could be argued that a context consists of relating resource systems, ecosystems and social, economic and political settings. Thirdly, the expected outcomes of the governance mechanisms and their fit with the original purpose are evaluated. For example, centralised governance structures are in general known to be effective for coordination of actions. Hence, they may have a high degree of fit for the purpose of immediate decision making during crisis situations.

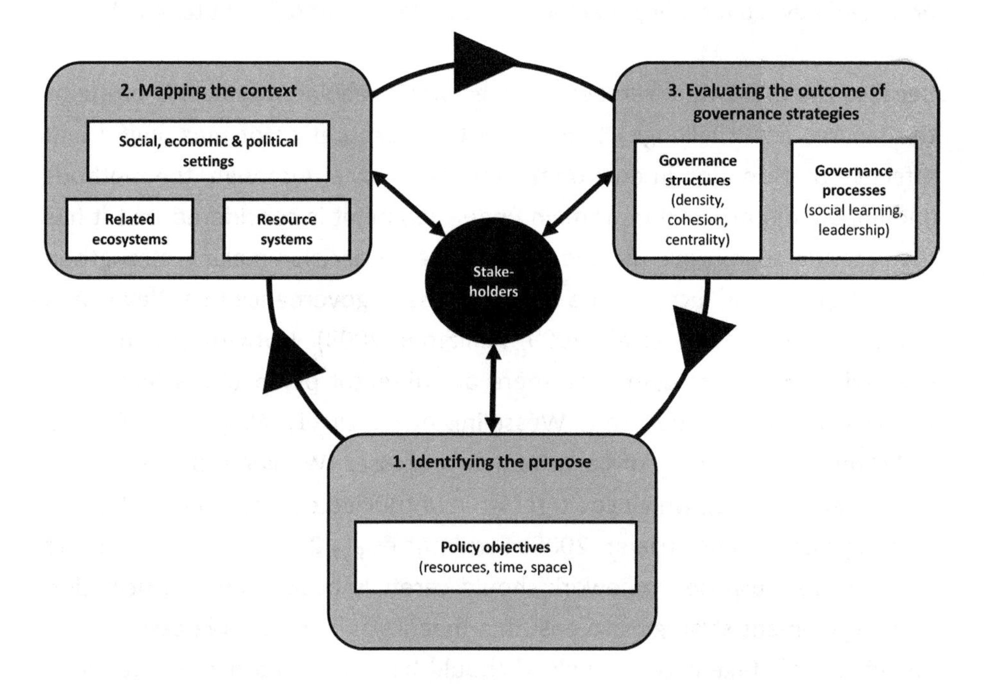

Figure 2.1 Three critical steps for diagnosing the fit-for-purpose of governance mechanisms: (1) identifying the purpose of governance; (2) mapping of the context; and (3) evaluating the outcome of governance mechanisms.

Stakeholders stand central in this model, because the outcome of the three steps depends on the mix of stakeholders within the assessment. Governance strategies arise from multi-stakeholder processes; thus, the purpose of governance mechanisms is also determined by multiple stakeholders. Their perspectives depend on their values, interests, knowledge and expectations. On the other hand, the purpose of governance strategies determines which actors have an interest and become stakeholders. By definition, stakeholders are operating in the context of governance. However, the context also shapes how stakeholders behave and interact with the physical environment. Because of the interdependencies between stakeholders and the purpose, context and fit of governance mechanisms, the fit-for-purpose governance framework requires a holistic approach that includes analysis of the purpose, context and fit from different stakeholder perspectives. Through taking a holistic perspective, the needs for new governance measures (i.e. the purpose), the legacy of existing governance mechanisms and challenges and opportunities (i.e. the context), and strengths and weaknesses of different proposed new governance mechanisms can be explored (i.e. outcomes).

Because the presented framework relies on stakeholder input, it is prone to the failures and challenges that relate to incorporation of meaningful and effective participation in environmental governance. Although, the methods and impact of participation remain under debate, it is considered that it has the potential to improve the knowledge base for decision making, strengthen public support and increase the effectiveness of governance (e.g. Newig and Fritsch, 2009; Paavola et al., 2009; Pellizzoni, 2003). Notwithstanding this, even within single assessments, there are different perspectives on the rationales for participation (e.g. Wesselink et al., 2011; Wright and Fritsch, 2011) and on the design of participation processes (Webler and Tuler, 2006; Webler et al., 2001), which could result in unfulfilled expectations and disappointing performance (Hajer, 2005; Turnhout et al., 2010). Hence, the users of the fit-for-purpose framework should carefully design their participation and engagement strategies to ensure a meaningful and reliable assessment. The choice of stakeholders involved should be based on a balance between economic efficiency, environmental effectiveness, equity and political legitimacy (Adger et al., 2003). Furthermore, the mix of actors involved in the assessment should encompass stakeholders at the operational, institutional and constitutional levels of governance, covering different governance functions (e.g. ownership and management functions) and consider all institutional rules that regulate the use and management of environmental resources (Paavola, 2007). This makes the use of the fit-for-purpose governance framework a timely process that relies on the user's ability to gain insight into these aspects of governance prior to or during the fit-for-purpose governance assessment.

2.4 First steps towards operationalisation of the fit-for-purpose governance framework

As described above, the purpose and contextual conditions depend on the values, beliefs and interests of the involved stakeholders. However, a review of adaptive governance literature (including the network management, leadership and social learning literatures) suggests that in general, different structures and processes have different strengths and weaknesses and may therefore in general be preferred in particular situations. In order to better understand governance outcomes, a review of network properties (i.e. governance structures and processes) has been conducted. Three key properties that describe network structure are identified from literature: density, cohesion and centrality of networks (see Table 2.1). The analysis suggests that properties under a given combination of purpose and contextual conditions provide different outcomes. For example, in immediate crisis situations such as flooding, timely and well coordinated responses are needed. In such a context, centralised network structures are likely to be more effective in coordinating action than in decentralised networks, where power is more distributed in the network (Ernstson et al., 2008). Using hierarchy, debate or conflicting actions may be avoided which may enable timely evacuation so that people can be saved from undesirable outcomes such as drowning. However, centralised coordination may, for example, cause legitimacy issues (Bodin and Crona, 2009; Ernstson et al., 2008) in adaptation to long-term structural changes such as water allocation in large-scale transboundary water systems. In this scenario, networks with a lower degree of centrality and cohesion (i.e. multiple communities) and a higher density (i.e. interconnectedness) may be more appropriate because they provide the diverse knowledge base that is needed for finding solutions to complex problems (e.g. Davidson-Hunt, 2006; Page, 2008).

In terms of governance processes, it is important to take note that complex adaptive systems evolve due to external pressure or self-organising interactions in networks. In adaptive governance literature, social learning and leadership are considered key processes on which self-organisation depends (e.g. Folke et al., 2005; Olsson et al., 2006). Therefore, I focus here on these processes rather than more traditional governance processes such as policy making, regulation, monitoring, compliance and enforcement, education and community engagement. Scholarship about social learning and recent literature about leadership both use complexity as a starting point. Both processes emerge from interaction between actors in a network. They are therefore not mutually exclusive. However, there are obvious differences of behaviour and outcomes between social learning and leadership processes (see Table 2). Social learning is a critical factor for increasing receptivity to new approaches or technologies (Jeffrey and Seaton, 2004), creating and nurturing adaptive governance (Pahl-Wostl, 2007) and system resilience (e.g. Folke, 2006), and establishing transitions of systems as a whole (e.g. Loorbach, 2010). Leadership acts as a catalyst to change in otherwise self-organising complex networks (Bodin and Crona, 2009; Olsson et al., 2006).

From a review of social learning literature (see Table 2.2), it can be concluded that social learning is in particular suitable to increase understanding of the nature, degree and implications of problems and alternatives, values and implications of solutions. The collaborative processes on which social learning are based can potentially create or increase trust and shared norms and values. However, social learning is a process that requires time and effort. Leadership, on the other hand, catalyses change through triggering and coordinating action and engaging new actors. Although it could be less time demanding, it requires individuals in the network with leadership skills at management or project levels and/or organisations who have the capability and are willing to take up leadership roles. Actions resulting from strong leadership are not necessarily supported by a cohesive network which may potentially lead to a lack of legitimate outcomes. It could be concluded that the different outcomes of social learning and leadership processes cause different levels of fit of the applied process with its purpose in a certain context. For example, social learning is not a logical process to apply when strongly coordinated action is desired to deal with an immediate crisis. However, the fit of network processes is not only determined by the physical and social context of the network, but also by the network structure in which processes take place. As I have described above, strongly centralised network structures are effective for solving relatively simple problems, but are less effective in dealing with complex issues. Such network structures rely on traditional models of transformative leadership, but are likely to be too formalised to allow for social learning.

Table 2.1 Governance Structures: Key properties of network structure

Property	Definition	Strengths	Weaknesses
Network density	The extent to which a network is interconnected. It can be calculated by the number of existing ties between network actors divided by the number of possible ties. In policy science, density is also referred to as interconnectedness See also (Bressers et al., 1994; Bressers and O'Toole Jr, 1998).	- A higher number of social ties enhances development of knowledge and understanding through increased exposure to information and new ideas (Granovetter, 1973). - A higher number of social ties between actors leads to more possibilities for collective action through increased possibilities for communication and, over time, potentially increased levels of reciprocity and trust (e.g. Axelrod, 1997; Hahn et al., 2006; Olsson et al., 2004a).	- Group effectiveness of collective action may decline at high densities (Oh et al., 2004). - Excessively high densities can lead to homogenisation of information and knowledge which, in turn, may lead to less efficient use of resources and reduced capacity to adapt to changing conditions (Bodin and Norberg, 2005; Little and McDonald, 2007; Ruef, 2002).
Network cohesion	The extent to which individuals, groups and organisations empathise with each others' objectives insofar as these are relevant to the policy issue (Bressers and O'Toole Jr, 1998). When there is limited cohesion, several communities can be distinguished in a network.	- The presence of multiple communities (lack of cohesion) may enhance the development of knowledge within communities by providing opportunities for high degrees of interaction between actors with similar interests, leading to increased capacity to transfer tacit knowledge (Reagans and McEvily, 2003), spread of attitudes and opinions (e.g. Faust et al., 2002; Padgett and Ansell, 1993; Porter et al., 2005). - The presence of multiple communities may	- A lack of cohesion may result in limited collaboration between communities when there is a lack of ties between these communities (Granovetter, 1973). - The presence of multiple communities may hinder transfer of tacit knowledge, because individuals have limited cognitive capacity and therefore are forced to be selective in keeping up their relationships with others (Bodin and Crona, 2009).

		contribute to the development of a diversity of knowledge by enabling various forms of knowledge to be developed in different communities, leading to increased adaptive capacity(e.g. Davidson-Hunt, 2006; Page, 2008).	
Centrality - of an actor - of a network	The extent to which an actor has a central position in a network. The extent to which there is variability of centrality between the actors in a network (Wasserman and Faust, 1994).	- By occupying central positions in a network, actors can influence others in networks and are better situated to access valuable information which can put them at an advantage (Burt, 1995, 2004; Degenne and Forsé, 1999) - Adoption of innovations is generally being diffused from cores of centralised actors to more loosely connected peripheral actors (Abrahamson and Rosenkopf, 1997). - Higher network centrality increases the ability to solve simple problems structures (Bodin et al., 2006; Leavitt, 1951). - Higher degrees of centrality are favoured for mobilisation and coordination of actions (Bodin et al., 2006).	- Actors have limited capacity to support and maintain network connections (Bodin and Crona, 2009). - Possibilities for action can be constrained when an actor feels obliged to please all its network neighbours (Frank and Yasumoto, 1998). - Complex problem solving requires more decentralised network structures structures (Leavitt, 1951) (Bodin et al., 2006). - Lower degrees of centrality may be favoured to engage a broad spectrum of stakeholders in order to resolve issues of complex governance processes in later phases (Bodin et al., 2006).

Table 2.2 Governance Processes; Overview of social learning and leadership

Process	Description	Strengths	Weaknesses
Social learning	Learning through interaction of individuals and/or communities (e.g. Folke, 2003; Pahl-Wostl et al., 2007). Three aspects of learning can be distinguished: research to enhance discovery and understanding, capacity building to enhance people's awareness and capabilities, and application to enhance practical outcomes (see also Senge and Scharmer, 2001).	- When applied in informal settings, social learning can facilitate the development of innovative solutions to existing problems by providing opportunities to explore new ideas, devising alternative designs, and testing policy (Gunderson, 1999; Olsson et al., 2006; van Herk et al., 2011a). As such, it plays an important role in connecting actors from different network communities (Olsson et al., 2004b; Olsson et al., 2006).	- Social learning is a time intensive process and requires the involvement of a range of stakeholders (van Herk et al., 2011a). - When social learning is organised in formal settings, members of social learning groups may feel scrutinised by their agencies or constituencies, resulting in limited freedom to learn from each other, think creatively and develop alternative solutions (Gunderson, 1999).
Leadership	Traditionally, scholarship has considered leadership in a transformational sense in which "leadership behaviours that inform and inspire followers to perform beyond expectations while transcending self-interest for the good of the organisation" (Avolio et al., 2009). More recently, complexity leadership theory has recognised that leadership is too complex to be described as only the act of individuals. From the perspective of complexity, leadership emerges from interaction between actors	- Transformational leadership can be characterised by persistence, enthusiasm, articulating inspiring vision, questioning the status quo, and providing inspiration and motivation to others (Bass, 1985, 1999). - From a complexity perspective, leadership enables rather than controls the future (Uhl-Bien et al., 2007). Enabling leaders recognise or create windows of opportunity (Olsson et al., 2006) to disrupt existing patterns of behaviour, encourage novelty, and make sense of emerging events for others (Boal and Schultz, 2007; Plowman et al., 2007). Furthermore, enabling leaders create	- Traditional forms of focused top-down leadership are usually ineffective in complex challenges, because they are not suited any more for the fast-paced, volatile context of the Knowledge Era (Marion and Uhl-Bien, 2001; Schneider and Somers, 2006).

	(Lichtenstein and Plowman, 2009; Uhl-Bien and Marion, 2009) and may occur as top-down, bottom-up and/or lateral processes (Avolio et al., 2009; Lichtenstein et al., 2006).	structures, rules, interactions, interdependencies, tension and culture (Marion, 2008).	

2.5 Concluding discussion

Adaptive governance is aiming to establish resilient systems. In the adaptive governance literature, it is argued that a mix of top-down and bottom-up management is well-placed to achieve this (see e.g. Berkes, 2002; Folke et al., 2005). Nelson et al (2007, p. 499) go one step further by stating that "the strong normative message from resilience research is that shared rights and responsibility for resource management (often known as co-management) and decentralisation are best suited to promoting resilience". Caution should be taken to avoid the conclusion that a multi-level governance approach alone is considered to be sufficient for establishing adaptive governance. Depending on the context and stakeholder needs, an adaptive approach can at different points in time include different purposes such as coordination of activities, generating new knowledge, and distributing knowledge. As identified above, different governance structures and processes have different strengths and weaknesses and are therefore to a varying degree appropriate for different purposes. By evaluating the effectiveness of existing and proposed governance mechanisms, the fit-for-purpose governance framework can be applied as both a descriptive and a prescriptive tool to operationalise adaptive governance. When applied to governance arrangements that are already established, this procedure provides information about necessary adjustments. For example, it could be used to evaluate the success of established adaptation policies, or to evaluate the effectiveness of governance arrangements to stimulate transitions to more sustainable or resilient environmental resource management. Furthermore, it provides a procedure that could be applied for prediction of the likely success of planned reform(s); for example assessing the ability of Australian urban water markets to efficiently allocate scarce water resources in an institutional context that is dominated by one water service provider and rigid health regulation.

The fit-for-purpose governance framework could also be considered a step back from adaptive governance, because it provides direction for conducting one particular evaluation rather than a continuous cycle of regular evaluations in time. Hence, it only provides a starting point for adaptive approaches. However, by making the incumbent uncertainties relating to adaptive governance explicit it makes policy makers aware about a need for deliberation when setting up or reforming governance arrangements. By doing so, it points their attention at adaptive governance principles through insights into ineffective or inappropriate governance activities. Meanwhile it provides a research agenda for scientists for assisting to put adaptive governance into practice. Based on a literature review, this chapter has shown that further work is needed for the development of practical tools for: 1) defining the purpose of governance and balancing interests, beliefs and values; 2) determining the relevance and impact of contextual conditions on different governance mechanisms; 3) determining the (expected) outcomes of governance mechanisms under different conditions.

The problem of fit is not new (e.g. Folke et al., 1998, 2007; Galaz et al., 2008; Young and Underdal, 1997). In particular, it is argued that matching governance with the dynamic characteristics of ecosystems and the inherent uncertainties related to (abrupt) change within both governance systems and ecosystems is challenging (Galaz et al., 2008). The fit of governance with its context depends on the temporal and spatial scales and the scope of institutions (Folke et al., 1998, 2007). In their words, "how does the scale (temporal, spatial, functional) of an institution relate to the ecosystem being managed, and does it affect the effectiveness and robustness of the institution?" (Folke et al., 2007, p. 2). The research about the problem of fit has attempted to enhance the fit through system evaluation (Ekstrom and Young, 2009), understanding different types of misfits (Galaz et al., 2008) and increasing understanding of adaptive (Olsson et al., 2007) and polycentric governance arrangements (Ostrom, 2010). In this chapter, I add to this context/fitness dialogue the importance of purpose of governance and the procedures in which policy practitioners work. By emphasising the policy practitioners' perspective, I aim to enrich the dialogue about the fitness of governance under different conditions.

However, I conclude that further research is needed to operationalise the concept of fit-for-purpose: because governance emerges from interaction between multiple stakeholders with multiple interests, beliefs and values, there are multiple perspectives about fit depending on individual interests and values. However, taking a holistic view and analysing the fit from different perspectives may give a good indication if there is a fit or not. Receptiveness of network actors to alternatives may indicate that there is a lack of fit in a certain system, because it indicates that an improvement could be achieved. Perhaps a stronger indicator for the fit-for-purpose of governance could be advocacy of network actors for alternatives. It is likely that advocacy is a stronger indication than receptiveness, because an advocate is committed to invest time, effort, and possibly capital and reputation to consider alternatives. Other indicators of lack of fit may be new scientific knowledge, disasters or community concern. Further work is needed to identify which indicators best determine the degree of fit in a specific context.

Hence, similar to the concept of adaptive governance, fit-for-purpose governance is not yet readily applicable in governance practice. The fit-for-purpose governance framework provides the ingredients for a diagnostic procedure, but lacks empirical evidence to show how the framework works in practice. However, it provides the basis for a new way of thinking to address impediments to the uptake of adaptive governance by using a procedure that has similarity with the predominant institutional arrangements of predict and control regimes in which most policy makers operate. As such, the fit-for-purpose governance framework provides an alternative starting point for developing the much sought-after guidance for policy and decision makers to evaluate the effectiveness of established governance arrangements and to predict the likelihood of success of institutional reform.

CHAPTER THREE

Configuring transformative governance to enhance resilient urban water systems

A pattern of effective governance.

This chapter addresses the 'when' question through application of the 'fit-for-purpose' governance procedure of chapter 2 on different stages of transformation of urban water systems that are adapting to drought. As such, a pattern of effective governance configurations during consecutive stages of system transformation is identified. This pattern can be used to provide guidance for urban water governance reform to policy makers and governance evaluators.

3. Configuring transformative governance to enhance resilient urban water systems

This chapter is adapted from:

Rijke, J., Farrelly, M., Brown, R., Zevenbergen, C. (2013) Configuring transformative governance to enhance resilient urban water systems. *Environmental Science and Policy* 25: 62-72.

Abstract

Governance reforms are required to establish adaptive and resilient urban water resource management that takes into account complexity, uncertainty and immediate and long term change. This chapter details the outcomes of a qualitative, social science research project, drawing on insights from Australian urban water practitioners (n=90) across three Australian cities to explore the effectiveness of governance reforms in the contemporary urban water context. The perceived effectiveness of current urban water governance strategies were assessed through the first application of a fit-for-purpose governance framework, which helps to assess whether the (anticipated) outcomes match the intended purposes of proposed and applied governance strategies. The research provides important insights regarding the need for a mix of centralised and decentralised, and formal and informal, governance approaches to support effective governance of water infrastructure during different stages of adapting to drought and transitioning to a water sensitive city that is resilient to immediate and gradual change. The research insights suggest that decentralised and informal governance approaches are particularly effective in early stages of transformation processes (i.e. adaptation and transition processes), whilst formal and centralised approaches become more effective during later stages of transformation. As such, I have identified a pattern of effective governance configurations during consecutive stages of transformation processes that could provide policy makers guidance in overcoming urban water governance challenges.

3.1 Introduction

In many places, water scarcity and uncertainty are forcing a re-think about the way governments manage their water resource management systems. As a result, approaches such as Integrated Water Resources Management (e.g. Biswas, 2004), Sustainable Water Resource Management (e.g. Loucks, 2000) and Water Sensitive Urban Design (WSUD; e.g. Wong and Brown, 2009) have (re-) gained prominence over the last decades to deliver water resource management systems that are adaptive to change and resilient to extremes. Collectively, these approaches are comprehensive systems approaches that involve multiple disciplines and stakeholder groups. Research related to these approaches has demonstrated that urban water reforms should result in resilient water resource management that explicitly takes into account complexity, uncertainty and immediate and long term change (Folke et al., 2005). Resilience provides capacity: (i) to absorb shocks while maintaining function (Holling, 1973); (ii) for renewal and reorganisation following disturbance (Gunderson and Holling, 2002); and (iii) for adaptation and learning (Folke, 2006; Gunderson, 1999; Olsson et al., 2006). Despite the availability of technologies and knowledge required to develop resilient water resource management systems, practical implementation remains slow (Harding, 2006; Mitchell, 2006). Developing resilient water resource management systems is more a governance issue than a technological issue as *"adaptation to climate change is limited by the values, perceptions, processes and power structures within society"* (Adger et al., 2009, p.349).

Meanwhile, there is insufficient prescription for transformative governance approaches that enhance resilient water systems (e.g. Loorbach, 2010; Rijke et al., 2012a). The purpose of prescription for transformative governance is twofold: 1) to enable adaptive capacity for establishing resilience (i.e. to enable adaptation); and 2) to transform existing systems into more resilient systems (i.e. to enable transitions). Creating effective prescription is complicated by the recognition that there are no blueprint solutions for good governance that operate successfully in all conditions and across all scales (Ostrom et al., 2007; Pahl-Wostl et al., 2010). 'Solutions' that have been widely implemented in the past, such as the privatisation of public services or decentralised management of natural resources, have a track record of repeated failure related to unanticipated outcomes (Acheson, 2006). However, several recent contributions have been useful for developing prescriptions for effective governance through guiding principles (e.g. Huntjens et al., 2012; Ostrom and Cox, 2010) and attributes of transformative governance (e.g. Farrelly et al., 2012; Loorbach, 2010; Pahl-Wostl et al., 2010; van de Meene et al., 2011). Whilst all these efforts provide general guidance for policy and decision makers to governance arrangements that enhance resilience, most of them fail to provide specific guidance for governance related to changing circumstances during transformation processes, with some recent exceptions (i.e. Adger et al., 2011; Herrfahrdt-Pähle and Pahl-Wostl, 2012; Olsson et al., 2006). Therefore, this article focuses on providing guidance for aligning effective governance strategies during different stages of transformation processes.

Urban water governance in three Australian cities is being drawn upon as Australian cities are facing highly variable and extreme climate conditions. Over the last decade, long-lasting drought interrupted by short periods of extreme rainfall have placed the traditional, large-scale water infrastructure under pressure regarding the security of water supplies and protecting cities from flooding. In response to such pressures, the concept of a water sensitive city (WSC) has emerged concurrently from the technical and social science fields (Brown et al., 2009b). A WSC is the outcome of WSUD processes, and is considered to be adaptive and resilient to broadscale change (i.e. demographic change, climate change and extreme weather conditions) and values water, promotes conservation and aims to improve liveability (Wong and Brown, 2009). Such a city would achieve this through planning for diverse and flexible water sources (e.g. dams, desalination, water grids and stormwater harvesting), incorporating WSUD for drought and flood mitigation, environmental protection and low carbon urban water services in the planning system, and enabling social and institutional capacity for sustainable water management (see also Wong and Brown, 2009).

Although technologies that make WSCs possible have been successfully demonstrated on a number of occasions (Farrelly and Brown, 2011), there remain significant institutional barriers to facilitating this paradigm shift in planning, design, operation and management of urban water systems including: a lack of understanding about urban water cycles; different interpretations of WSUD; WSUD values are not firmly embedded in the water and development sectors; limited skills and competencies to apply WSUD; a fragmented urban water space; a limiting regulatory environment for technological innovation; and, ineffective leadership (see also Brown and Farrelly, 2009b; Pahl-Wostl, 2007). Because of these challenges, there are no examples of a city which has fully transformed into a WSC. Therefore it should be noted that empirical evidence that a WSC is indeed adaptive to change and resilient to extremes is not available. However, the terminology of WSCs and WSUD is being used in Australia to assist cities in adapting to a persistent drought (2001-2009) and climate change. For example, the terminology has been adopted as a policy objective in the National Water Initiative (National Water Commission, 2011) and in the South Australian Government's *Water for Good* strategy (Office for Water Security, 2010). Hence, the overall objective of this chapter is to identify the patterns of governance configurations that are likely to be most effective as a system transforms to a WSC that is posited to be more adaptive to change.

3.2 Social-ecological and socio-technical perspectives on governance

Over the last decades, several efforts have been made to better align the physical domain with the concept of governance. From a social-ecological perspective, in which social systems continuously interact with ecosystems, the concept of resilience emerged in the 1970s, introducing the notions of dynamic equilibria and multi-stable states (Holling, 1973). Building on the social-ecological perspective and the concept of resilience, adaptive governance emerged as a way of governing by anticipating long-term change (i.e. climate change, population growth), responding to immediate shock events (i.e. drought, flooding) and recovering from such events (see also Folke et al., 2005). Since the late 1990s, a socio-technical perspective has emerged from technology and innovation studies. It examines how societal systems - including culture, politics, institutions and economics - and technical systems co-evolve over time. It focuses upon transitions, which are long-term nonlinear processes (25-50 years) that result in structural changes in the way a society or a subsystem of society (e.g. water management, energy supply) operate (Rotmans et al., 2001). Governance to establish transitions, often referred to as transition management, aims at influencing interactions between the dominant 'regime' (meso level) with its societal 'landscape' (macro level) and 'niches' (micro level) where innovation occurs, so that these innovations become mainstream (Berkhout et al., 2004; Geels, 2002; Rip and Kemp, 1998).

Social-ecological systems and socio-technical systems are considered to behave as complex adaptive systems; they change as a result from self-organisation and external pressure (de Haan, 2006; Scheffer, 2009). Therefore, unsurprisingly, adaptive governance and transition governance share several characteristics and challenges (Foxon et al., 2009; Smith and Stirling, 2010). Although adaptive governance focuses on the ability to maintain system functions under changing conditions whilst transition governance focuses on the ability to steer structural system change, the social-ecological and socio-technical system literatures suggest that adaptive cycles and transitions follow a comparable process of subsequent stages and activities (see Table 3.1). Furthermore, the adaptive governance and the transition governance scholarships are both focused on learning through, for example, "shadow networks" (Olsson et al., 2006), learning and action alliances (van Herk et al., 2011a) and transition arenas (Loorbach, 2010). To anticipate change and retain the ability to adjust adaptive and/or transformative strategies to changing drivers and problems, continuous monitoring and evaluation, and iterative adjustment of governance practices are required (Loorbach, 2007; Voß et al., 2006).

Table 3.1 Typical activities during transitions and adaptive cycles

Transition stage (socio-technical system)	Adaptive cycle phase (social-ecological system)	Typical activities
Pre-development	-	Network formation, experimentation, learning.
Take-off	Re-organisation/renewal	Response to a crisis or establishment of a policy decision.
Acceleration	Growth/exploitation	Increasing implementation of innovation.
Stabilisation	Conservation	Regulation and legislation to establish the status quo
-	Collapse/release	Losing faith, searching for new/alternative solutions

(adapted from de Haan and Rotmans, 2011; Folke, 2006; Geels and Schot, 2007; Gunderson, 1999; Holling and Gunderson, 2002; Olsson et al., 2006; Rotmans et al., 2001; Walker et al., 2004)

From the socio-technical and the social-ecological perspectives, multi-level (or polycentric) governance is considered crucial for enhancing resilient water management (Huitema et al., 2009; van de Meene et al., 2011). In a multi-level governance system, decision making is dispersed across multiple centres of authority (Hooghe and Marks, 2003). As such, it is the outcome of interaction between public sector agencies, private sector organisations and the community. Multi-level governance enables knowledge exchange and mutual adjustment of governance at different levels and sectors of governance (Agrawal, 2003) and potentially leads to synergetic effects (Ostrom and Cox, 2010) that enable more adaptive governance regimes (Armitage et al., 2007). Multi-level governance can vary between different degrees of centrality. In its extreme form, centralised governance can be conceptualised as a hierarchy. Hierarchies enable powerful actors operating at the top of the hierarchy to access valuable information and control action(s) by making others accountable (Degenne and Forsé, 1999; Kjær, 2004). Extreme decentralised governance can be conceptualised as markets or networks. Market governance uses private sector management principles (Hood, 1991), such as market pricing and competition, to allocate resources efficiently and empower citizens (Pierre and Peters, 2000). Network governance is the outcome of self-organisation resulting from interaction between a broad spectrum of stakeholders (Bodin and Crona, 2009; Olsson et al., 2006) and is founded upon reciprocity and consensus (Kjær, 2004). Furthermore, multi-level governance relies on a mix of formal institutions and informal networks (Olsson et al., 2006; Tompkins and Adger, 2004). Formal institutions typically include legislative and regulative frameworks, whereas informal networks, also referred to as "shadow networks" (Olsson et al., 2006), play an important role in connecting actors, learning, knowledge management, and accessing resources and support (Gunderson, 1999; Olsson et al., 2006). The interaction amongst actors in informal networks largely takes place outside the scrutiny of formal forums (i.e. regulatory, policy and planning processes) (see also Gunderson, 1999; Olsson et al., 2006).

3.3 Research approach

The objective of this chapter is to identify the patterns of governance configurations that are likely to be most effective as a system transforms to a

WSC that is posited to be more adaptive to change. Drawing on the current understanding regarding transformative governance outlined in section 3.2, three case studies have been undertaken for comparative purposes. Although each of the cities has very different institutional infrastructure and existing governance strategies, the three selected cities (Sydney, Melbourne and Adelaide) have all experienced the same macro-scale pressures related to persistent drought (e.g. Brown and Clarke, 2007; Brown and Farrelly, 2009b; Daniels, 2010), suggesting that each city potentially represents a different stage of transformation. The differences between the case studies provides an opportunity to identify patterns, if any, regarding the applied governance strategies across the three cities for addressing the macro-scale pressure.

For each of the cases, the recently developed fit-for-purpose governance framework was used to provide a snapshot of the effectiveness of urban water governance approaches in a particular phase of an adaptive response to a long-term drought and a transition towards water sensitive cities (Rijke et al., 2012a). This framework is specifically designed to guide the evaluation of the effectiveness of urban water governance strategies to enable transformation and consists of three steps: (1) identifying the purpose of governance; (2) mapping of the context; and (3) evaluating the outcome of governance mechanisms. In this article, I follow the policy objectives of the Australian Federal Government's National Water Initiative (National Water Commission, 2011) and choose the purpose of governance as follows: to effectively enable adaptation to drought by transforming to WSCs. A particular emphasis of this article focuses on the governance context of stormwater management as this has been identified as a critical factor in establishing WSCs in the Australian context (Wong and Brown, 2009). The governance context was mapped in each of the cases using the guidance from the socio-technical and social-ecological system theories regarding the various stages of transformation (Table 3.1).

This chapter focuses on the fit during different stages of system transformation. As outlined in Table 1, these stages are being shaped by interactive processes. The fit of the governance strategies that were employed by the stakeholders in each of the three cases was evaluated using stakeholder perceptions about the strengths and weaknesses as a proxy for the effectiveness of governance. Subsequently, I have assessed the effectiveness of governance during different stages of transformation by matching the effectiveness of the proxy with the activities that are typically performed during these respective stages of transformation. The proxy is based on the argument that the effectiveness of interactive processes depends on how satisfied actors are with these processes, because actors judge whether the objectives are met (Edelenbos and Klijn, 2006; Klijn and Teisman., 1997). Explicitly discussing the strengths and weaknesses of individual governance strategies with the actors involved enhances, therefore, the understanding of the effectiveness of such strategies during different stages of transformation. However, Folke et al (1998, 2007) argue that the fit of governance with its context also depends on spatial scales of institutional mandates and the ecosystem they are managing. Accordingly, I also note that the scale of infrastructure can relate to the degree of centralisation of governance (Elzen and Wieczorek, 2005; Lieberherr, 2011). This chapter focuses on governance in the context of system transformation, thus temporal dimensions of governance reforms are the main focus and the assessment of spatial fit is beyond the scope of this chapter.

Table 3.2 Overview of respondents in first interview round (July - September 2010)

	Adelaide	Melbourne	Sydney	Total
State government	17	5	11	33
Local government	5	3	8	16
Water utilities	6	8	4	18
Private sector	2	4	3	9
Professional associations	3	2	3	8
Other	3	1	2	6
Total	36	23	31	90

The analysis of the case studies is based on insights that were derived from a set of 90 face-to-face interviews and the analysis of policy reports, legislation, regulation and media documentation. The interviews were semi-structured and covered four general topics: 1) the (historical) context of urban water governance; 2) current activities to achieve a WSC; 3) the strengths and limitations of these activities. Interviewees represented a range of different disciplines and organisations, which included key decision makers and individuals in senior advisory roles. The interviewees included both insiders of urban water governance in each of the three cities (e.g. water practitioners, government representatives and community representatives) and individuals who were more disconnected from the cases but who had a overview about urban water management across Australia (e.g. individuals in professional associations, politics, consultancy, science; see Table 3.2). The interviewees were selected to represent different perspectives regarding the purpose of urban water governance (e.g. supporting vs. opposing WSUD, advocates of economic vs. social/environmental valuation of natural assets, advocates of centralised vs. decentralised governance) to ensure to capture potentially conflicting governance strategies.

Following the above first stage of data analysis, an extensive validation process was undertaken to test the research findings. Firstly, policy reports, legislation, regulation and media documentation were collated to support and/or contradict practitioner interpretations. Secondly, the findings related to the effectiveness of governance strategies were compared to the scientific literature about adaptive governance, multi-level governance, network management, leadership and social learning. Thirdly, the findings were discussed with the actors in the three cities through: i) validation workshops with actors (total n=81) in each city and ii) interviews (n=15) with 20 representatives of key stakeholder groups which were previously interviewed. Fourthly, the findings about the case studies were compared through discussions with individuals with an overview about urban water governance in Australia (n=12).

The analyses of the fit of governance in the three case studies were collated in to develop a pattern of governance activities that fit the various stages of transformation. Based on the results about the fit of governance in each of the cases, generalised configurations of fit-for-purpose governance are determined for various transformation stages. Subsequently, a pattern of these fit-for-purpose governance configurations is developed.

3.4 Australian cities responding to drought

Traditionally, Australian urban water management is largely the responsibility of State Governments, which have predominantly relied on a highly centralised 'big pipes-in big pipes-out' infrastructure (see, e.g. Newman, 2001).

Each of the cities has its own specific urban water context. For example, with regard to stormwater management, Adelaide has local governments playing a strong leadership role (since the early 1990s) with regards to innovation around stormwater harvesting and reuse technologies; in Melbourne, the state water authority has a long history with demonstration of stormwater treatment technologies and has mandated WSUD in its State planning regulations; whereas in Sydney, there has been limited State Government attention to stormwater management since 2006, but increasing attention and action from key local governments.

In terms of WUSD adoption, Melbourne is further ahead than the other two cities as its State Government was the first, and to date only, Australian State Government to mandate WSUD principles for potable water conservation, water reuse and stormwater management for new residential developments through the amended Victorian Planning Provisions (see also Brown and Clarke, 2007). Indeed, respondents in Adelaide and Sydney almost unanimously referred to Melbourne as an example in terms of establishing WSUD. Since the early 1990s, the Melbourne water authority, water retailers, state government, certain local governments, the private sector and research institutes have collaborated to further develop the WSUD concept through its implementation in practice (Brown and Clarke, 2007; Farrelly and Brown, 2011). Despite Melbourne expanding its WSUD practices, this approach is yet to become mainstream, due to the slow uptake of WSUD principles in practice (Farrelly and Brown, 2011; Morison, 2010). Therefore, I argue that Melbourne is currently in the growth phase of the adaptive cycle and approaching stabilisation of its transition to a WSC.

In Adelaide, the drought has forced the South Australian government to rethink the way it manages its water resources (see also Daniels, 2010; Rijke et al., 2011). At the peak of the drought, supplies from the river Murray stopped and the city nearly ran out of water. This situation triggered a centralised governance response to water scarcity that extends the existing dominance of highly centralised, large-scale technologies such as desalination, wastewater recycling schemes, and improving the efficiency of existing systems through better interconnectivity and introducing market trading schemes (see also Office for Water Security, 2010). In addition, and at the time of interviewing, Adelaide had also made large-scale investments in stormwater harvesting schemes and had begun to institutionalise WSUD (see also Rijke et al., 2011). These investments were preceded by local governments experimenting with stormwater harvesting and reuse schemes since the early 1990s (Daniels, 2010; Rijke et al., 2011). As an interviewed senior policy maker (State government) stated: *"It's been a very fractured space... ...but I think we're at the point now where all the forces are converging and the institutional arrangements are now going to support the development of the central urban water effort."* Overall, I argue that Adelaide is currently in the re-organisation phase of the adaptive cycle and post take-off in its transition to a WSC.

Although Sydney has the largest per capita water storage system in Australia, during the drought, storage volumes approached 30% (in early 2007) (NSW Office of Water, 2010). According to interviewees, Sydney's urban water sector is focused primarily upon expanding its large scale, centralised water supply practices. Following the completion of the State sponsored Stormwater Trust in 2006, WSUD has received considerably less attention and state-based funds have ceased to exist for stormwater management. However, there are a number of local government initiatives which are aimed at improving stormwater quality. Overall, there appears to be a significant disconnect between local and state government agencies – which has subsequently resulted in gaps between urban water policy and practice, particularly in relation to stormwater management. The State Government sees its primary focus as water supply and acknowledges the important role local governments and the catchment management authorities play as key partners in addressing water quality management. Over the course of interviewing, a common complaint amongst local government respondents suggested "*Nobody knows who to speak to in State Government*" and local government representatives expressed their "*disappointment in leadership from the State government*", stating that "*it feels we're never a stakeholder*". Despite local government initiatives which challenge the status quo, their results are not adopted or supported by the State Government, suggesting that Sydney remains in a conservation phase and is not (yet) transitioning to a WSC.

Based on document analysis and the interview data, Table 3.3 provides an overview of the main governance initiatives that were employed after the peak of the drought in 2006 in order to a) secure water supplies and b) protect waterway health (i.e. to achieve water sensitive cities). The research data show that these are centralised to different degrees, depending on the scale level of infrastructure systems and the involvement of federal, state, regional or local government agencies. Furthermore, the research data reveals how urban water governance in the three cities relies on a mix of formal institutions and informal networks. Based on the research findings, I suggest that urban water governance in the three cities is, albeit to different degrees, a mostly centralised affair and largely reliant on formal institutions (Figure 3.1). For example, governance in Adelaide is, to a large extent, coordinated by the Department for Water which was established in 2010 as the leading urban water policy organisation. In addition, there exists a strong informal network of planners, engineers and policy makers across all stakeholder groups which provides, with fluctuating intensity, input to governance processes. In Melbourne, a similar, but slightly less connected informal network was identified; although, in comparison to Adelaide, there is less centralised coordination of urban water governance. Instead, leadership is more distributed amongst key stakeholder groups, such as the State water authority, water retailers, local governments and research institutions. Governance in Sydney relies almost exclusively on formal institutions. Although informal networks were identified, they mostly operate within formalised coalitions of local governments around stormwater management.

Table 3.3 Main governance initiatives to enable water sensitive cities

Main governance initiatives (2006 – 2011)	centralised	decentralised	formal	informal
ADELAIDE				
– Establishment of Commissioner for Water Security to prepare holistic strategy for securing water supplies (*Water for Good*)	x		x	
– Establishment of Department for Water as new leading urban water policy agency	x		x	
– Large scale investment in large scale stormwater harvesting and wastewater recycling schemes	x	x	x	
– Development of desalination plant	x		x	
– Announcement of a state wide mandate for WSUD	x		x	
– Development of urban water markets		x	x	
– Leadership through informal networks		x		x
– Establishment of De Goyder research institute to support State Government in securing water supplies	x		x	

MELBOURNE				
– Demonstrating change through a large number of experiments and pilot projects (see also Brown and Clarke, 2007; Farrelly and Brown, 2011)		x	x	x
– Mandate for WSUD through the Victorian Planning Provisions	x		x	
– Large scale investment in large scale stormwater harvesting and wastewater recycling schemes	x	x	x	
– Development of desalination plant	x		x	
– Development of urban water markets (attention diminished after State election)		x	x	
– Centralised efforts to build networks and competencies through the Clearwater programme	x		x	x
– Distributed leadership across the water authority, water retailers, local governments, State government		x	x	
– Leadership through informal networks		x		x
SYDNEY				
– Policy framework for securing water resources by a mix of supply and demand measures: Metropolitian Water Plan 2010	x		x	
– Decentralised efforts to institutionalise WSUD through Local Environmental Plans and Building Control Plans and BASIX certificates		x	x	
– Local learning and action alliances for experimentation and joint learning		x	x	x
– Development of desalination plant	x		x	
– Large scale investment in large scale wastewater recycling schemes	x		x	
– Establishment of the Water Industry Competition Act 2006 to enable urban water markets		x	x	

3.5 Perceived strengths and weaknesses of applied governance strategies

3.5.1 Respondents' perceptions

To determine the level of fit of the governance approaches that are applied in each case study city, interviewees were asked to identify the perceived strengths and weaknesses of the aforementioned governance approaches (Table 3.3). Table 3.4 presents the aggregated (validated) responses regarding water practitioner perceptions about the strengths and weaknesses of the identified governance initiatives. The results have been aggregated as they were similar across all three cases for the categories centralised governance, decentralised governance, formal institutions and informal networks.

3.5.2 Discussion

The findings presented in Table 3.4 are similar to what the scientific literature describes about centralised/decentralised governance, formal institutions and informal networks.

Table 3.4 Perceived strengths and weaknesses of applied governance strategies across all three cases

	Perceived strengths	Perceived weaknesses
Centralised governance	- Potential of relatively rapid decision making - Effective coordination of resources and activities and synergies - Centralised regulation potentially provides a bottom line and a fair playing field	- Low capacity to solve complex problems - Risk of illegitimate and unfair outcomes - Causes others to walk away from responsibilities
Decentralised governance	- Involvement of multiple actors and disciplines: access to larger knowledge base	- Ineffective use of resources (knowledge, capital, fte) - Risk of limited knowledge sharing
Formal institutions	- Binding and irreversible	- Silo mentality - Low adaptive capacity
Informal networks	- Building trust - Exploring problems and solutions - Incubation of innovative ideas - High adaptive capacity - High degrees of tacit knowledge	- Fluctuating levels of interaction - Vulnerable to losing knowledge - Difficulties to tap from informal networks

Regarding centralised governance, the scientific literature describes how hierarchies are favoured to solve simple problems (Leavitt, 1951) and mobilise and coordinate action (Bodin et al., 2006), but have a low capacity to solve complex problems and are at risk of resulting in illegitimate and unfair outcomes (Marion, 2008; Schneider and Somers, 2006). Decentralised governance approaches are advocated in the policy science, transition management and adaptive governance scholarship for addressing complex problems through involving a wide diversity of knowledge, social learning and collaborative leadership (e.g. Bodin and Crona, 2009; Bodin et al., 2006; Olsson et al., 2006). Although decentralised approaches are founded upon reciprocity and consensus (Kjær, 2004), they can also lead to institutional fragmentation, and poses challenges for accountability (Kjær, 2004) and capacity in managing and maintaining network connections (Bodin and Crona, 2009).

Governance traditionally relies on formal institutions where legislative frameworks set out the rules and responsibilities of organisations involved (Kjær, 2004; Pierre and Peters, 2000). Formal agreements are often binding and difficult to reverse (Kjær, 2004; Pierre and Peters, 2000). As a result, they generally have a low adaptive capacity which often is being aggravated by silo mentality between and within organisations (Brown, 2008b). In addition, informal networks of practitioners and decision makers play an important role in building trust between disciplines and organisations, because they operate outside the scrutiny of organisational (formal) mandates (Gunderson, 1999; Olsson et al., 2006). As a result, they provide a consistent driver for exploring and understanding problems and solutions and act as an incubator for innovative ideas. For example, the development and implementation of WSUD in Melbourne can be attributed to a network of 'champions' that operate across a range of organisations within the urban water sector (see Brown and Clarke, 2007; Taylor et al., 2011). Van Herk and colleagues (2011b) argue that if networked partnerships are cultivated in the long term, they can stimulate collaborative learning and promote transitions. The identified informal networks were considered to be highly adaptive and contain high degrees of tacit knowledge (e.g. about local geophysical conditions, community, relationships, history). However, they remain vulnerable to losing such knowledge, as they operate voluntarily and are thus prone to network members moving in and out generating fluctuating levels of interactions (see also Gunderson, 1999).

3.6 Configuring transformative governance

3.6.1 Early stages of transformation

The early stages of transformation processes can be classified as the pre-development and take-off stages in socio-technical systems theory and as the collapse and re-organisation stages in the social-ecological systems theory (Table 3.1). As described in Section 3.4, Adelaide is leaving the early transformation stage whilst Sydney is struggling to enter it.

During this early stage of transformation to a WSC, problems with the traditional approach to urban water management have become apparent and individuals started to re-consider traditional 'solutions' in both cities. Experimentation, learning and network formation are playing an important role during these early stages of system transformation, because these activities generate the trust in new technologies and collaborations (Section 3.2). As described in Section 5, decentralised governance and informal networks are generally considered most appropriate to support these activities. Indeed, I have observed that decentralised (collaborative) leadership and policy making often occurs in conjunction with formalised, learning alliances that provide "safe spaces" for informal collaboration and trust building. For example, in Sydney the Cooks River Sustainability Initiative and the Georges River Urban Sustainability Initiative were established by local governments for collaborative learning purposes. By involving and using multiple sources of knowledge these initiatives and labelling the activities as demonstration, these activities generated trust in WSUD technologies and stimulated collaboration between various disciplines, such as planners and engineers to reduce the risk of undesirable outcomes (see also Bos and Brown, 2012). Also, in all three cases, I observed that informal networks of professionals were instrumental in cross-disciplinary and inter-organisational exchange of knowledge and experiences (e.g. engineering, urban planning, natural resources management) and collaborate in advocacy activities for the uptake of WSUD technologies.

A comparison between the cases of Adelaide and Sydney shows that successful demonstration of technologies needs to be followed up by a centralised policy decision in order to proceed into the next stage of transformation. In Adelaide, respondents were almost unanimously positive about centralised efforts to: (i) adopt a vision for WSCs, (ii) provide economic incentives, (iii) establish state wide policy framework for adopting WSUD (similar results were found in the other cases). In response to the drought, a Commissioner for Water Security was appointed in Adelaide to establish a coherent water management strategy which secured future water supplies. The strategy, *Water for Good,* included among others, the designation of appropriate resources for future water supplies and the announcement of a WSUD mandate that would, according to interviewees, not have been possible without a directive approach. As one local government representative reflected: *"I think in retrospect the existence of the Office of Water Security probably made the preparation of that plan a lot more efficient than it would have otherwise been. Because otherwise you would have had to develop it by consensus and I think, within the State, from my experience that's always difficult with different departments having different priorities and whilst the Cabinet obviously has to rule off in the end on the State plan, it's obviously a lot easier if you have appointed one person or one office to actually do that coordinator work, so I think that was a master stroke in the end."* As a result, the Federal, State and local governments have jointly invested $150 million in stormwater technologies in Adelaide in 2009 to make the city increasingly water sensitive (Wong et al., 2009). On the contrary, in Sydney, a centralised governance response remains pending, resulting in a limiting regulatory environment that hampers take-off of a transition towards a WSC. However, it should be noted that the sense of urgency to adapt was higher in Adelaide than it was in Sydney: Whilst Adelaide's main water supply was nearly cut off due to extremely low water levels in the River Murray, Sydney's water reservoirs remained approximately 30% full at the peak of the drought (NSW Office of Water, 2010).

3.6.2 Mid stages of transformation

The mid stages of transformation processes can be classified as the acceleration stage and the growth stage in, respectively, the socio-technical and social-ecological systems theories (Table 3.1). In Section 3.4 it was described that Adelaide is entering and Melbourne is leaving these stages. For example in Adelaide, the establishment of the *Water for Good* strategy and the AU$150 million investments in stormwater harvesting and reuse demarcate the start of the mid stage of transformation in which WSUD is shifting from demonstration to mainstream practice. Although centralised policy decisions were in both cases unanimously perceived as highly effective, there were interviewees who remained sceptical (predominantly decentralised stakeholders such as local governments). These respondents argued that centralised approaches are inappropriate for solving complex problems, such as developing an integrated urban water management strategy, for they tend to overlook stakeholder interests and contextual knowledge related to particular areas, technologies, stakeholder relationships and history. Therefore, they argue that such approaches are at risk of overly relying on traditionally powerful perspectives and actors, such as engineering and the water authorities, and overlooking other relevant stakeholders such as urban planners and ecologists.

Another concern raised was that centralised leadership may lead to certain stakeholders walking away from their responsibilities. As an example, following the establishment of the Department of Water as the leading urban water policy agency in Adelaide, the State planning agency took a step back from preparing the State's WSUD mandate, which has led to difficulties in securing the implementation of the mandate. This example illustrates also that reliance on formal institutions can lead to a silo mentality between different departments within and between organisations. This challenge was commonly raised amongst all interviewed stakeholder groups across all three cases. For example, Sydney local government respondents revealed their frustrations about "*the rigid attitude*" of regulators to adopting stormwater innovations, and in Melbourne a respondent commented on the simplistic "*one-dimensional view of the world*". Particularly during the validation workshops, the view that such institutional fragmentation hampers innovation and the development and results in low levels of adaptive capacity was commonly shared. Particularly the respondents who were during the early stages of transformation involved in the informal networks and learning alliances suggested that these stayed important during mid stages of transformation in order to build trust and stimulate collaborative learning.

However, based on water practitioner perceptions across all three cases, I note that decentralised approaches are not always effective as they may result in ineffective coordination and use of resources, as barriers need to be overcome multiple times in different (local government) organisations, and there is a risk of limited knowledge sharing (e.g. between local governments or learning alliances). In Adelaide, the partners in the *Waterproofing Ade-*

laide projects recognised an opportunity for State government coordination of multiple stormwater harvesting and reuse project that were being implemented simultaneously by different local governments. According to the involved interviewees, this centralised governance approach was selected to enable greater knowledge sharing within and across organisations through collecting, collating, synthesising and distributing project outcomes and lessons learned and, as such, effectively make use of the resources available (i.e. financial and human capital, knowledge).

3.6.3 Late stages of transformation

The late stages of transformation processes can be classified as the stabilisation stage and the conservation stage in, respectively, the socio-technical and social-ecological systems theories (Table 3.1).

With the establishment of the state-wide mandate for adopting WSUD in new residential developments, Melbourne has commenced entering this stage. The Melbourne respondents commonly shared a perception that centralised (state-wide) regulation creates an new bottom line and a level playing field for all stakeholders. Respondents from local government and the water authority often commented that a mandate for WSUD removes inconsistencies in land development processes through performance standards of WSUD technologies. In Adelaide, where the development of a mandate for WSUD in all new developments is announced in the Water for Good strategy, respondents (primarily within local government) identified that establishing state-wide regulation for WSUD is largely dependent on political leadership. In Melbourne, collaborative leadership of executives and project officers across water authority, land developers, local governments and science has ultimately convinced the state government to adopt WSUD in the planning regulations (see also Brown and Clarke, 2007; Taylor et al., 2011).

Although Melbourne has adopted regulation that requires adoption for WSUD in all new residential developments, it has not yet transformed into a WSC. The values of WSUD are not yet firmly embedded in everyday practice (Brown and Farrelly, 2009b). Whilst some municipalities are fully embracing WSUD, others are staying behind (Morison, 2010). Interviewed practitioners and policy makers have commented that sufficient awareness about value, capacity, competences and guidelines is required to implement regulation in practice. In Melbourne, capacity building activities are coordinated through the Clearwater programme, which is mainly funded by the water authority. Although respondents considered central coordination of these activities to be resource intensive, they suggested that centralised capacity building efforts enable greater knowledge sharing within and across organisations through collecting, collating, synthesising and distributing project outcomes and lessons learned.

3.6.4 A pattern of effective governance configurations

Linking the insights about the effectiveness of governance with the activities that have taken place during different phases of transformation provides a pattern of effective governance configurations during different stages of transformation. The research findings that are described in Section 3.6.1-3.6.3 are collated and summarised in Table 3.5.

3.6.5 Discussion

The pattern that is described in Table 5 provides policy makers and governance evaluators assistance in linking effective governance configurations during different stages of transformation. As such, the effectiveness of governance configurations relates to the ability to transform and not to the effectiveness of a WSC itself. As empirical data about the performance of a WSC on a city scale are currently unavailable, it can therefore not be concluded that the identified pattern provides desirable bio-physical output (e.g. secure water supplies, flood protection, good water quality). However, the research findings presented enable policy makers to accelerate transformation towards WSCs, which could in turn provide insights in whether the concept of a WSC is indeed resilient and adaptive to change.

Whilst the social-ecological systems theory does not specify transformation pathways other than the adaptive cycle (i.e. collapse, re-organisation, growth, conservation), the socio-technical systems theory distinguishes between various forms of transformation. For example, Berkhout et al (2004) make a distinction between planned and unplanned system change based on internal or external resources (i.e. resources coming from inside/outside the dominant regime). In addition, de Haan and Rotmans (2011) distinguish transition patterns that consist of top-down (*"reconstellation"*) and bottom-up (*"empowerment"*) system change and internally induced system change (*"adaptation"*). Arguing that no transition is completely planned upfront, but all transitions involve some degree of coordination through the alignment of visions and activities of various actors, Geels and Schot (2007) have identified that transition pathways can be distinguished by the timing (e.g. macro pressure before/after innovations are fully developed) and nature (i.e. reinforcing/disruptive relationships) of interactions between actors. In this chapter, the governance pattern identified relates to *reinforcing* interaction to *coordinate* (as far as possible) a transformation to a WSC through combining top-down and bottom-up governance approaches to adapt to changing circumstances.

Table 3.5 Effective governance under different circumstances

Transition stage	Adaptive cycle phase	Typical activities	Effective governance approaches
Pre-development	-	Network formation, experimentation, learning.	*Decentralised and informal*: to establish and nurture new relationships and test innovations
Take-off	Re-organisation / renewal	Response to a crisis or establishment of a policy decision.	*Hybrid*: formal policy decision to catalyse and/or coordinate activities, and informal and decentralised learning to further test innovations
Acceleration	Growth / exploitation	Increasing implementation of innovation.	*Hybrid*: centralised policy to enable activities, decentralised implementation, informal network to distribute tacit knowledge, coordinated capacity building to create synergies and avoid inefficient use of resources.
Stabilisation	Conservation	Regulation and legislation to establish the status quo	*Centralised and formal*: to adjust or establish legislative frameworks and coordinated capacity building to convince and enable laggards to adopt innovative approaches and safeguard a new status quo.
-	Collapse / release	Losing faith, searching for new/alternative solutions	*Decentralised and informal*: to search for alternative solutions and share experiences.

This research has drawn upon three Australian cities adapting to persistent drought. As such, the findings regarding the fit of governance approaches and the pattern of effective governance during various transformation stages are tailored to the Australian urban water context. However, urban water management in other parts of the world, such as Europe and North America are undergoing similar transformation (Hoyer et al., 2011). The relevance of the research insights presented stretch therefore beyond urban water in Australian. However, I strongly recommend to design governance reform after assessing urban water governance approaches using the procedure of the 'fit-for-purpose framework', as it reduces the risk for selecting inappropriate governance approaches by explicitly taking into account context and purpose of governance reform (Rijke et al., 2012a). It is important to remember, however, that as the fit-for-purpose governance assessment only provides a snapshot in time of urban water governance approaches, guidance for improving governance approaches will need to be regularly reviewed.

3.7 Conclusion

Governance of adaptation is a matter of continuous learning and making timely decisions. This chapter describes that different stages of the transition process towards WSCs favour different configurations of centralised/decentralised and formal/informal governance. The early transformation stages typically involve the invention and testing of new technologies and processes and the formation of informal networks. Decentralised and informal governance approaches are favourable to enable such activities (Section 5). In addition, centralised policy decisions can further stimulate the uptake of innovations and the coordination of decentralised projects can enhance the efficient use of resources by creating synergies through sharing relevant knowledge. During the mid transformation stages, the importance of informal networks remains to maintain connections and distribute tacit knowledge across different institutions and disciplines. Also, coordinated capacity building has been proven effective to create synergies and avoid inefficient use of resources. During late transformation stages centralised and formal approaches were considered effective to adjust or establish legislative frameworks and coordinate capacity building to convince and enable laggards to adopt innovative approaches and safeguard a new status quo.

By taking a combined view, the research findings presented are enriching the bodies of literature on adaptive governance and transition governance. Both bodies of literature have identified hybrid multi-level governance approaches that balance between centralised control and bottom up approaches for learning e.g. (Huntjens et al., 2012; van de Meene et al., 2011) to be important for successful adaptation and establishing socio-technical transitions. In addition, my findings provide preliminary guidance for the timing of different governance configurations. As such, the identified pattern of effective governance configurations can be used to provide guidance for urban water governance reform to policy makers and governance evaluators. However, I recommend further research regarding appropriate guidance for achieving and organising hybrid governance approaches.

CHAPTER FOUR

Room for the River: Delivering integrated river basin management in the Netherlands

Criteria for change.

This chapter addresses the question of why transformational processes are sometimes being hampered. Through application of an existing 'transitions governance' framework, a set of criteria for establishing structural system change (i.e. a transition) is tested for the context of river flood protection in the Netherlands. These criteria can be used as a checklist for policy and decision makers to establish systemic transformations, such as transitions and system-wide adaptation.

4. Room for the River: Delivering integrated river basin management in the Netherlands

This chapter is adapted from:

Rijke, J., van Herk, S., Zevenbergen, C., Ashley, R. (2012) Room for the River: Delivering integrated river basin management in the Netherlands. *International Journal of River Basin Management* 10(4): 369-382.

Abstract

This chapter describes how the governance arrangements of the 2.2 billion[1] Euro water safety programme Room for the River are enabling a transition towards integrated river basin management in the Netherlands. I observe that in terms of integrating multiple objectives and spatial scales, the programme design and multi-level governance processes in the programme have enabled the establishment of integrated plans and designs. I conclude that Room for the River plays an important role in a transition to integrated river basin management in the Netherlands through practical implementation of the strategic policy vision for integrated water management. Also, through application of a mixed centralised-decentralised governance approach, the programme has tackled governance pitfalls related to centralised planning approaches that previously impeded integrated water management. Although several of the governance lessons of the programme are being adopted by for example the Delta Programme, I have identified a risk that continuity of the newly introduced governance approach may be lost when the Room for the River program is completed in 2015.

[1] This chapter mentions a total budget of €2.2 billion instead of €2.4 billion , because it is written before the inclusion of the IJsseldelta project in the Room for the River programme.

4.1 Introduction

Traditionally in the Netherlands, water management was seen purely as a matter of civil engineering and aimed at controlling nature. Saeijs (1991, p. 245) illustrated this by writing: *"God created man, but the Dutch created their own land"*. Over the last thousand years, this attitude towards water management has resulted in gradual development in the washlands of the country's rivers that has reduced room for them and required repeated heightening of flood defenses. However, several policy scientists argue that, in common with many other countries, a transition has been taking place in flood risk management in the Netherlands since the 1970s. In this transition, the traditional sectoral engineering approach to flood risk management is gradually being replaced by an integrated approach that incorporates various disciplines such as water management, spatial planning and ecology (e.g. van der Brugge et al., 2005; van Stokkom et al., 2005; Wiering and Arts, 2006). A similar change is taking place in flood risk management in for example, Europe and North America (Warner et al., 2013).

The first time that the 'control paradigm' was challenged and adapted to include ecological values was during the national public debate that led to a significant alteration of the original construction plan of the Eastern Scheldt storm surge barrier in 1974. Irreversible ecological damage of the salt water environment that would be caused by the closing of the Eastern Scheldt estuary was avoided through the construction of moveable panels that would only be closed under extreme circumstances (Knoester et al., 1984). Later in the 1980s, the technocratic control paradigm was further challenged by the emergence of the concept of *"integrated water management"* that originally aimed to avoid conflicts between different uses of water resources through improved coordination (Saeijs, 1991). Although these new principles were already applied in the 1970s and the 1980s, it was not until after the 1993 and 1995 near-miss river floods that a new policy window opened for the implementation of integrated water management (van der Brugge et al., 2005; van Stokkom et al., 2005; Wolsink, 2006).

In 1995, extreme river water levels nearly caused dike breaches and led to the evacuation of 250,000 people and 1 million cattle. This created enhanced awareness amongst the public, politicians, public administration and water professionals that nature cannot be controlled and that new ways of managing rivers was required; i.e. through creating more space for rivers to discharge their flows. As an ad hoc response, a new policy line, the Room for the River Directive, was developed by the Dutch government (ten Heuvelhof et al., 2007). This Directive established that water should be considered as a structuring principle for spatial development (Oostdam et al., 2000; Valk and Wolsink, 2001). However, as Wolsink (2006) points out, spatial developments are still mostly being driven by economic and social priorities. In line with this, the report of the Dutch governmental advisory Commission 'Water Management 21st Century' recommended mutual adjustment of water and spatial conditions rather than water as the leading structuring principle (CW21, 2000).

The Room for the River Directive resulted in the approval of the governmental decision for the Room for the River programme (PKB Ruimte voor de Rivier) by the Dutch Senate in December 2006. The 2.2 billion Euro Room for the River programme, began the detailed design phase in 2006 and is scheduled for completion by 2015. It has a dual objective of: 1) improving safety against flooding of riverine areas of the Rivers Rhine, Meuse, Waal, IJssel and Lek by accommodating a discharge capacity of 16.000m^3/s; 2) contributing to the improvement of the spatial quality of the riverine area. At the start of the programme, a set of 39 locations was selected for giving more room for the rivers through, for example, flood by-passes, excavation of flood plains, dike relocation, and lowering of groynes (Figure 4.1).

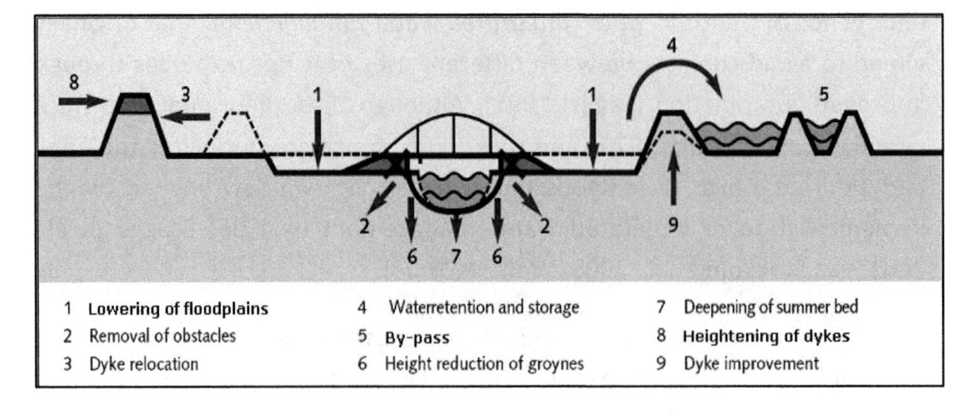

Figure 4.1 Measures that are applied in Room for the River (Source: Room for the River Programme Office)

Although this meant the commencement of large-scale implementation of an integrated water management approach, several assessments of the uptake of integrated water management concluded, at the time of commencing the plan study phase of the Room for the River programme (2006), that the transition towards integrated water management was not complete (van der Brugge et al., 2005; Wiering and Arts, 2006; Wolsink, 2006). For example, Van der Brugge et al (2005) comment that, at the time of writing, there was still a considerable gap between strategic policy visions and practical implementation. Furthermore, Wolsink (2006) pointed out that governance pitfalls related to centralised planning cultures (lack of participation, lack of consideration of local identity in planning decisions) impeded integrated water management. Similarly, Wiering and Arts (2006) concluded that, at the time, it was too early to speak of *"deep institutional change"*, because although the traditional water institutions were opening up to other disciplines they were maintaining their power positions.

However, the Room for the River programme has adopted a new (multi-level) governance approach in which government agencies in different disciplines (e.g. water safety, planning, agriculture, nature) and at national, regional and local levels are actively collaborating (van den Brink, 2009). The programme uses a mix of centralised (national) steering/decentralised (regional) decision making processes (see also ten Heuvelhof et al., 2007). The decision frameworks for establishing improved water safety and spatial quality are set by the national Government, whilst the plans and designs are formulated and decisions taken by local and regional stakeholders in 39 regional projects. The national government has established a central programme office to manage and monitor progress, evaluate quality of designs and facilitate the regional projects through guidelines, providing expert knowledge, community building, and where needed, applying political pressure. This approach provided the opportunity for decentralised governments to link local issues such as new developments and the development of natural and recreational areas with the water safety agenda (Hulsker et al., 2011; van Twist et al., 2011b).

At present (August 2012), most of the initial 39 regional projects within the Room for the River programme have completed their planning phase and entered the realisation phase (PDR, 2011b). Meanwhile, Room for the River is considered an *"exemplary project"* for adopting new governance approaches by the Ministry of Infrastructure and Environment and Rijkswaterstaat (the executive arm of the Dutch Ministry of Infrastructure and the Environment, which is responsible for the design, construction, management and maintenance of the main infrastructure facilities in the Netherlands) and (van den Brink, 2009, p. 15). For example, the recently established Delta Programme (2009-2015) is using Room for the River as an example for governance and developing integrated strategies. The Delta Programme is currently preparing Delta Decisions for securing water safety (against flooding) and fresh water supplies. These Delta Decisions will be ready in 2015 and will be implemented according to the Delta Act that provides a continuous funding stream of 1 billion Euro per year into a Delta Fund from 2012 and beyond. Hence, the lessons from Room for the River have potentially major implications for future water management in the Netherlands.

These lessons also have international relevance, as the concepts of making space for rivers and new multi-level governance approaches are also being adopted by other countries. For example, the concept of making space for rivers is being applied in countries, such as France, Germany, Hungary, Romania, the UK and the USA (DEFRA, 2007; Opperman et al., 2009; Warner et al., 2013). Although the motivations for the concept of making space for rivers vary in these countries, the implications in terms of governance are similar: multi-level governance approaches are considered to be required for overcoming controversies between various actors involved (Warner et al., 2013). Similarly, many others advocate a multi-level governance approach for integrated water management. (e.g. Huitema et al., 2009; Huntjens et al., 2012; Pahl-Wostl et al., 2010; van de Meene et al., 2011). However, the scale of taking a structured and integrated view to flood protection as being applied in the Netherlands is not being matched elsewhere (Warner et al., 2013). In this light, the Netherlands is collaborating intensively with countries such as the USA, Vietnam, Bangladesh and Indonesia to adopt integrated approaches to water management (see also Zevenbergen et al., 2012).

In the light of these new ideas and initiatives and the leading position of the Netherlands, I have re-assessed the transition towards integrated river basin management in the Netherlands in this chapter by examining to what extent the governance arrangements of the Room for the River programme influence this shift. In order to do this, I first identify how the governance arrangements of Room for the River are enabling integrated plans and designs. Subsequently, I discuss the experiences and lessons learnt from Room for the River and how these are being adopted in the Delta Programme.

4.2 Theoretical framework

4.2.1 Integrated river basin management

The concept of integrated river basin management is derived from integrated water resources management. The term 'integrated water resource management' is interpreted differently by many (see also Biswas, 2004). Overall, integrated water resource management is a comprehensive approach (Mitchell, 2005). For example, it is defined as *"a process which promotes the coordinated development and management of water, land and related resources, in order to maximise the resultant economic and social welfare in an equitable manner without compromising the sustainability of vital ecosystems"* (Global Water Partnership, 2000). It refers to an holistic approach to the whole water cycle, including natural flows (i.e. precipitation, stormwater, groundwater, surface water and evapotranspiration) and man-made flows (e.g. drinking water, sewage and recycled water). It also relates to different functionalities of water systems, such as providing water safety, transport capacity, water security, and healthy ecosystems. Similar to integrated water resource management, integrated river basin management is a comprehensive and coordinated approach. The main difference is that it focuses explicitly on river basins.

From the scientific literature, three different perspectives on integrated river basin management can be distinguished. Firstly, integration is about alignment and balancing of *multiple objectives.* For river basin management, objectives such as providing safety, transport capacity, opportunities for recreation, enabling nature, facilitating economics, safeguarding aesthetics and water quality play an important role (e.g. Downs et al., 1991; Opperman et al., 2009; Saeijs, 1991). Integrated river basin management particularly takes into account the interplay between water and land use functions (Hooijer et al., 2004; Moss, 2004). Secondly, an integrated approach is a systems approach that includes all relevant *spatial scales* (see also Adger et al., 2005b;

Zevenbergen et al., 2008): systems as a whole and parts of systems such as components and elements (see van Herk et al., 2006). Relevant spatial scales for river basins are for example: catchment and sub-catchment scales (Jaspers, 2003; Savenije, 2009); and international, national, regional and local scales. Thirdly, comprehension of short and long term *time scales* in order to balance short and long term costs and benefits and anticipate (potential) future change (see also Adger et al., 2005b; Zevenbergen et al., 2008). For example, the definition of the Global Water Partnership for integrated water resource management that is quoted above includes the word sustainability, which is about meeting present needs without compromising the ability to meet future needs (see also Brundtland, 1987).

Summarising the above, I define integrated river basin management as a comprehensive water management approach that aligns multiple objectives in a river basin across different spatial scales and temporal dimensions.

4.2.2 Governance of change

"Water governance refers to the range of political, social, economic and administrative systems that are in place to regulate development and management of water resources and provisions of water services at different levels of society" – Global Water Partnership (2002)

The above definition provides an indication regarding the meaning of water governance. However, governance is a concept rooted in the social sciences and as such is defined and interpreted in many different ways (for an overview of definitions and interpretations, see e.g. Kjær, 2004; Rhodes, 1996). Governance incorporates both processes and structures required for steering and managing parts of societies (Kooiman, 1993; Pierre and Peters, 2000). As a process, governance refers to managing networks, markets, hierarchies or communities (Kjær, 2004; Rhodes, 1996), whereas governance as structure refers to the institutional design of patterns and mechanisms in which social order is generated and reproduced (Voß, 2007). Taking a combined view, governance can be considered as comprising three mutually reinforcing elements: policy (problems and solutions), polity (rules and structures), and politics (interaction and process) (Voß and Bornemann, 2011). Governance is also the outcome of interaction among multiple actors from different sectors with different levels of authority (Agrawal, 2003). As such, governance relies on institutions consisting of cognitive (dominant knowledge, thinking and skills), normative (culture, values and leadership) and regulative components (administration, rules and systems) that mutually influence practice (Scott, 2001).

A transition is a structural change in the way a society or a subsystem of society (e.g. water management, energy supply, agriculture) operates, and can be described as a long-term non-linear process (25-50 years) that results from a co-evolution of cultural, institutional, economic, ecological and technological processes and developments on various scale levels (Rotmans et al., 2001). As such, transitions are structural changes of practices, institutions and culture. Managing transitions requires continuous influence and adjustment in governance systems (Foxon et al., 2009; Loorbach, 2007; Smith and Stirling, 2010).

The scholarship that focuses on governance of change (transition management) has emerged over the last 10 to 15 years and is still developing significantly. A common critique is that it lacks prescription for effectively establishing change. However, several attempts have been made to overcome this knowledge gap, including a framework for transition management (Loorbach, 2010), principles for institutional design (Huntjens et al., 2012), and a procedure for evaluating the effectiveness of proposed approaches (Rijke et al., 2012a). A 'transitions governance framework' was recently developed from a series of studies in the urban water sector in Australia (Farrelly et al., 2012; see Appendix A). This framework consists of eight socio-institutional factors that are considered to have the capacity to influence existing and future governance approaches, and hence the ability to adopt new practices (see Table 4.1).

The 'transition governance framework' distinguishes between structural factors and process factors. The structural factors are relatively stable over long timeframes, but remain subject to reinterpretation through the process factors, which can adapt more readily to changing circumstances over shorter timeframes. However, these processes will ultimately influence, but will also be guided by, the core structural attributes.

Table 4.1 **Operational factors supporting transition governance (Farrelly et al., 2012; see Appendix A)**

Operational factors	Sub-components
STRUCTURE	
Narrative, metaphor and image (e.g. a clear vision)	Storyline that invokes a need for change Visual connection to problems and potential solutions
Regulatory and compliance agenda	Objectives and mechanisms (markets, legislative rules and education) Performance targets Monitoring, enforcement and evaluation
Economic justification	Demonstrated business case Appropriate allocation/evaluation of all social and environmental costs and benefits (monetary and non-monetary)
Policy & planning frameworks & institutional design	Define the scope of the policy Highlight the distribution and trade-offs of costs and benefits Legislation, administrative organisational arrangements Dedicated funding streams
PROCESS	
Leadership	Distributed network leadership (policy, operational, private sector, science, community and political) Organisational leadership Positional and personal leadership characteristics
Capacity building and demonstration	Creating awareness about problems and solutions Build confidence in approach, technology and practice Develop new skills and competencies across the sector Creating informal incentives to apply and replicate leanings
Public engagement and behavior change	Understanding existing community drivers Informing and engaging with the community Encouraging behavior change amongst community members
Research and partnerships with policy/practice	Science partnerships: co-constructing science, policy and practice agendas for evidence-based decision-making

4.3 Methodology

The aim of this chapter is to describe to what extent the governance arrangements of the Room for the River programme are enabling a transition towards integrated river management in the Netherlands. I have based my findings on a document analysis, a series of face-to-face interviews (n = 55; see Table 4.2) and a quantitative survey (n=151). All interviews covered similar topics including: 1) the connection between water safety and spatial quality; 2) the output of the programme in terms of integrated solutions; 3) the organisation of the programme / the management of the regional projects; 4) the uniqueness of Room for the River; and 5) lessons for the future. In addition, the results of a quantitative survey were used to confirm the interview responses about the output of Room for the River for when the design phase of nearly all plans in the program is completed. In total, there were 151 survey respondents (48 from the Room for the River programme office, 10 from other parts of Rijkswaterstaat, 10 from the Government Ministries involved, 11 from Provinces, 22 from waterboards, 36 from municipalities, 7 from the private sector and 7 other , such as scientists, community groups). Survey respondents included individuals working for the Programme Directorate, regional project teams, policy makers and Delta Programme staff, as well as executive decision makers at the national, regional and local governments.

Table 4.2 **Data collection (interviews) and validation**

Method	Description	Respondents /participants
Interviews	Central programme office	14
	Central programme office (interface management with projects)	7
	Regional projects (team members and politicians)	17
	Senior advisors to the programme	8
	National policy makers	6
	Management of follow-up programmes (Delta Programme)	3
Validation	Validation workshops (2x)	33
	Observing stakeholder management training sessions (3x)	45
	Observing political conferences (2x)	Approx. 220
	Observing knowledge symposium for Rijkswaterstaat professionals	Approx. 150

Based on the data, the extent to which the output (i.e. the designs and plans) of Room for the River can be considered integrated (in terms of objectives, spatial scales and temporal dimensions) has been evaluated. The interview and questionnaire responses about the governance of the Room for the River programme have been structured according to the 'transition governance framework' (Section 4.2.1.2) in order to analyze the extent to which Room for the River's programme design (input) and governance processes enabled integrated river basin management. In addition, I have assessed to what extent Room for the River's political and economic context has been influencing integrated output and outcome. Subsequently, I discuss how the lessons from Room for the River are being adopted in the Delta Programme and how translation could be improved.

Validation of the findings occurred through a workshop with officials of various government agencies and a validation workshop with a user panel comprising senior policy advisors; observations of three stakeholder management training sessions for three individual regional project teams; and observations from two political conferences for national, regional and local decision makers from the Room for the River project areas and one conference for professionals within Rijkswaterstaat (see Table 4.2).

4.4 Research Findings

4.4.1 Governance factors for establishing integrated river basin management outcomes

4.4.1.1 Vision

Soon after the occurrence of the extreme water levels in the Dutch river systems in 1993 and 1995, a new perspective rapidly became dominant amongst politicians, water managers, spatial planners and scientists that nature cannot be controlled. It was decided that high water discharges should not lead to higher water levels, because this would lead to increased damage during flooding. As such, it was decided that flood safety should be enhanced by giving more room for the rivers rather than heightening of flood defences (e.g. Hooijer et al., 2004; van der Brugge et al., 2005). The aforementioned groups, and particularly the individuals who were involved with the programme design of Room for the River, were aware that this required close collaboration between water management, spatial planning, and other disciplines such as ecology and landscape architecture. Furthermore, partly due to previous experiences from large railway infrastructure projects (see Section 4.3.2), a vision became 'common-currency' that traditional top-down governance of the programme would no longer be appropriate. Instead, close collaboration between governments at various levels was needed. This led to a steering philosophy of 'controlled trust' rather than top down governance.

In line with the advice of the Commission Elverding, which urged large infrastructure projects in the Netherlands in 2008 to apply improved planning processes for *"faster and better"* results (Commissie Elverding, 2008), Room for the River aimed to deliver the proposed measures before 2015 through stable decisions throughout the project. To avoid delays, the vision sought to involve politicians and non-governmental stakeholders early in the planning process to establish commitment and support; and to deliberately create overlap between separate planning stages (initiation, planning, realisation) to generate input early in the planning process from actors responsible for regulation, operation and maintenance in order to establish realistic plans and designs of good quality.

4.4.1.2 Policy framework

The initiation phase (2000-2006) of Room for the River worked towards the Room for the River Policy Decision (PKB Ruimte voor de Rivier) (see also ten Heuvelhof et al., 2007). The decision was agreed by the Ministries of Public Works, Spatial Planning, and Agriculture and Environment, regional waterboards and Provinces, and the Association of Municipalities. A shared document set out the integrated vision by setting a double objective: 1) improve safety against flooding by accommodating a discharge capacity of 16.000m^3/s in Lobith (where the Rhine crosses the German border); and 2) contribute to the improvement of the spatial quality of the riverine area. It stated that water safety is the leading objective. In addition, the PKB selected 39 locations for measures to be implemented and provided general ideas for the types of measure at these locations.

Furthermore, the PKB documented the procedures for the planning and realisation phases of the programme and the roles and responsibilities of the stakeholders. It described the principle that decentralised steering and execution of tasks should be applied where possible. To this end, a central programme office was established at Rijkswaterstaat to monitor progress, quality of plans, and achievement of objectives. This also documented how the steering philosophy of 'controlled trust' should be executed.

After the national government approved the PKB, Room for the River officially started in December 2006. By documenting the vision, objectives, procedures, roles and responsibilities in a document that was supported and co-signed by all the levels of government involved, the PKB provided guidance and a point of reference for the later phases of Room for the River. Similarly, for each individual project, the roles and responsibilities of the project partners in the planning stage were determined in 'agreements for cooperation' (samenwerkingsovereenkomsten). And for the realisation stage, 'realisation agreements' (realisatieovereenkomsten) described the quality, budget, time, market approach, project control methodology and risk distribution between region and Rijkswaterstaat. However, additional directions were also developed whilst the programme was underway (Section 4.3.1.6).

4.4.1.3 Economic justification

Budget-wise, Room for the River is primarily a water safety programme. The PKB states that the programme budget is allocated for improving water safety. However, because this budget also provides for integration of the measures in their local contexts, to alleviate resistance from local communities and secure maintenance after the measures are implemented. For example, in the dike relocation project in Deventer, the programme budget provides for the construction of an earth mound on which a farmer will build an organic dairy farm that will take responsibility for maintenance of the new flood plain. The programme office has since 13 December 2010, been obliged to indicate in their progress reports (Section 4.3.1.4) which part of the budget is allocated for increasing water safety and which part for other policy areas. In the 18th progress report that was sent to Parliament, the programme office estimates the total cost for achieving water safety objectives and integration of the designs in their existing environments as € 2169.5 million. In addition, third parties funded (e.g. municipalities, provinces) € 80 million for spatial developments (PDR, 2011a).

4.4.1.4 Regulation and compliance

It is the task of the central programme office to make sure that Room for the River's objectives are achieved on time and within budget. A proactive justification cycle that consists of monitoring, facilitation and justification was set up by the programme office to fulfil this role (Figure 4.2).

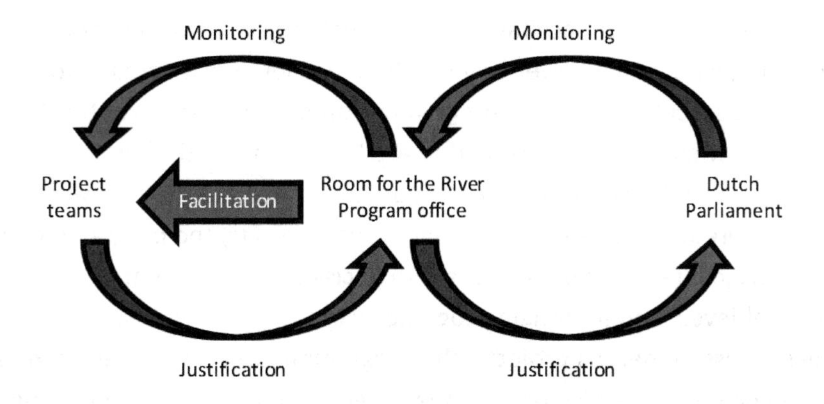

Table 4.3 Room for the River's justification cycle

According to the milestone management procedure that is described in the PKB (Section 4.3.1.2), the project teams need to follow a pre-set design process and deliver products (e.g. alternative design options, preferred designs, final designs and supplementary material) to the programme office. It is the core task of the programme office to evaluate the quality of these products in terms of hydraulic performance, spatial quality, legal procedures, geotechnics, integrated design, budgets and risk management. Every 6 months the programme office is required to send a progress report about the programme as a whole to the Dutch Parliament.

The programme office continuously monitors the project teams through 'river branch managers', who have regular interaction with project teams in order to achieve the desired quality and progress within the programme's boundary conditions. These 'river branch managers' form the interface between the programme office and the project teams and establish the link with the people who conduct the evaluations. This continuous connection between the project teams and the programme office facilitates the pre-assessment of the attainment or otherwise of the milestones before progress is submitted for approval. This provides the opportunity for the programme office to take timely action to facilitate improvements where the programme is not on track. Furthermore, an independent 'Q-team' of experts from multiple disciplines visited all projects at least three times to assess and give advice about the spatial quality of the plans.

The programme office facilitates the regional project teams in various ways. When project teams or 'river branch managers' identify a demand for expertise in a particular area, the programme office provides a central base of expert knowledge for all aspects that are being evaluated. If the programme office does not have the required expertise, it gives advice to project teams about where to acquire it. Also for individual projects, the programme office assists project teams in discussions with decision makers at national, regional and local levels and in bringing together various stakeholders. For common issues across numerous projects, the programme office prepares handbooks and guidelines to assist project teams and avoid them 're-inventing the wheel'. For example, handbooks are prepared for topics for spatial quality, underground cables and pipes and risk assessment. If the programme-wide realisation of projects is inhibited by existing policy and legislation, the programme office discusses this with national policy makers and legislators. This has resulted in, for example, changed policy arrangements for geotechnics transportation outside and inside dike-rings (see also van Herk et al., 2012c).

4.4.1.5 Leadership

According to the stakeholders involved, collective leadership is one of the cornerstones of Room for the River. I have observed that the governmental collaboration as documented in the PKB, is also taking place in the planning and realisation phases of the programme. On an organisational level, formal arrangements such as political covenants for initiation of projects (bestuursovereenkomst), collaboration agreements for the planning phases (samenwerkingsovereenkomst) and realisation agreements for the realisation phases (realisatieovereenkomst) define the collaboration by setting out the ambitions and responsibilities. In all projects within Room for the River, the national government is the client. In the initiation stage, the programme office executes the assignment. During the planning phase, each project has been assigned an 'initiator' (i.e. province, waterboard or municipality) who has formal responsibility to complete the plans. During the realisation phase, a 'realisator' (in most cases Rijkswaterstaat) has been responsible for acquiring permissions, tendering and contracting private parties for the implementation of the plans.

With regard to individuals, different networks were identified in which actors collaborate to deliver the projects. At the level of project officers', formal and informal forms of collaboration can be identified. Each of the project teams is shaped according to a standardised model consisting of a project manager (budget, time), risk manager, stakeholder manager, technical manager and a contract manager. As such, different competencies are organised within the project teams. Outside teams it is possible to identify informal networks of individuals that fulfil similar roles in different projects in which common issues are discussed. However, interaction in these networks has occurred mostly occasionally and ad hoc. At the level of decision makers, representatives from the organisations involved interact to build momentum and develop organisational commitment and legitimacy in the community. Also, I have identified several cases where informal interactions in networks of decision makers have played an important role to develop solutions when problems arose at a project level.

4.4.1.6 Capacity building and demonstration

Three factors that enhanced the programme's capacity to achieve its objectives were identified (see also van Herk et al., 2012a). Firstly, the programme office deliberately learned from the initial projects, such as depoldering of the Noordwaard and the Overdiepse Polder, the dike relocation at Westenholte and the excavation of the floodplains in the Schellener and Oldeneler Buitenwaarden. As a result, it was able to improve, for example, interface management between programme office and project teams, the configuration of project teams, milestone management procedures and deliverables, and technical guidance of projects on topics like underground cables and pipes, geotechnics transportation and risk assessment. Individuals in the project teams and in the programme office described how learning from these early projects would not have been possible without the centralised knowledge management and quality control by the programme office.

Secondly, continuous adaptation of governance processes to changing circumstances occurred within the programme office to be able to deal appropriately with various issues at different stages of the programme. For example, initially the task of the programme office was to monitor progress and quality of the work in the 39 project teams. When the programme office recognised that certain expertise (e.g. hydraulic, geotechnical, legal) was inadequate in the project teams, it played a more facilitating role. When many projects shifted into the realisation phase, a shortcoming of required expertise (e.g. market approach, tendering, logistics, litigation) was identified. As a result, the programme office enhanced its interface for each individual project with a senior staff member from its knowledge department and its project control department to support the 'river branch manager' who previously managed the interfacing task alone. Staff members of the programme office considered the pro-active justification cycle (Figure 4.2) as instrumental for signalling potential problems and solutions.

Thirdly, the programme office built capacity in the professionals involved through training for stakeholder management. Moreover, it actively stimulated community building through organising various network events that are tailored for particular roles, such as decision makers, project managers, stakeholder managers and risk managers. Experiences are shared at these events, and problems and potential solutions discussed. This has resulted in a Room for the River community and informal networks in which lessons are shared.

4.4.1.7 Public engagement

In Room for the River, governance is not only about collaboration between different government levels and agencies, but also about early involvement with the community. One of the reasons for giving the lead to regional governments was that these governments would more easily achieve community support and/or invoke less resistance because they are considered to know the local community better than the national Rijkswaterstaat. Furthermore, Room for the River has a, for Rijkswaterstaat's standards, unusually large communication office to proactively inform the communities involved and promote the programme. In this chapter I do not consider the RvR programme in terms of community contentment and support, as the chapter is concerned with professional practice and governmental transitions. However, several issues with regard to public engagement became apparent during the research. Examples identified include, inter alia, that where community engagement takes place too early in the planning process, it could lead to excessive expectations by the community and frustrations when it later appears that these expectations cannot be fulfilled. If engagement occurs too late (when there is no opportunity for adjustment of designs), it also leads to frustration and an increased risk of legal procedures. Also, an evaluation of the planning and design processes revealed that dominance of a small group in participative processes could result in sub-optimal designs that do not represent the common good (Hulsker et al., 2011).

4.4.1.8 Research

As outlined in Table 4.1, this Section focuses on science partnerships in which science, policy and practice agendas are co-constructed for evidence-based decision-making and enabling transitions. According to the interviewees, scientific research has not played an active role in the programme to support learning processes at a project level, for example through continuous reflection, the provision of scientific expertise, or as a platform to bring together different disciplines. Instead, research institutes were involved only to audit and validate models and calculations. At the start of the programme, the hydraulic model was validated and standardised for all 39 projects by a leading water research institute. Otherwise, the involvement of science occurred mostly in an ad hoc way. For example, to provide a second opinion or independent advice within individual projects.

However, at a programme level, (scientific) research has been used in the programme as an instrument to evaluate the processes, progress and outcomes. After completion of the PKB and after completion of the planning phase of the majority of the projects, processes and outcomes of Room for the River were evaluated by teams of policy scientists and consultants (ten Heuvelhof et al., 2007; van Twist et al., 2011b). In addition, a team of consultants has evaluated the design processes and output in terms of spatial quality of the programme (Hulsker et al., 2011). The findings of these evaluations have been used by the programme office to confirm its' decisions and to adjust governance arrangements where necessary.

4.4.2 The influence of contextual factors

The context of the programme has influenced the programme design at the start of Room for the River and its governance during the execution. As described above, the extreme water levels of 1993 and 1995 triggered the decision for Room for the River. However, the vision of making more room for rivers was on several occasions questioned by engineers, economists and politicians. For example, opposing engineers argued that the traditional approach of dike improvement has been successful for a long time and that river widening was not a proven method to provide better or cheaper solutions (e.g. Vrijling, 2008). Further criticism was that a cost-benefit assessment of the approximately 700 measures that were initially considered in Room for the River's initiation phase suggested that the programme was necessary and beneficial, but that creating more room for rivers was not the cheapest option for all river branches (Ebregt et al., 2005; Eijgenraam, 2005). However, at that time, political decision makers were not convinced by these criticisms and decided for realisation of the programme because they argued that river widening would add more value to the river area (e.g. economic, nature, recreation) and was a more effective flood risk measure (as failure of higher dikes would result in more water in polders and thus more damage).

In the current context of economic crisis and changed political priorities, the importance of transparent and cost-effective solutions is emphasised. As a consequence, the water safety objective of Room for the River has gained more weight as being the leading objective compared with the second objective of contributing to spatial quality (see also van Twist et al., 2011b). This expresses itself in the idea that nature is considered a luxury in contemporary Dutch politics and the loss of e.g. agricultural land should be reduced to a minimum. Because of this, the interpretation of spatial quality has changed within the programme from a focus on nature to a focus on agriculture.

Earlier events that took place prior to Room for the River have influenced the programme. Top-down governance approaches in delivery of the railway projects Betuweroute and the High Speed Line provoked resistance amongst local communities and politicians and led to delays which created widespread community scepticism about large infrastructure projects. This was one of the reasons for applying a combined centralised-decentralised governance approach within Room for the River. Following 2001, the response to the outbreak of foot-and-mouth disease that led to preventive elimination of cattle caused suspicion in regional communities towards the national government. In, for example, the flood bypass project Veessen-Wapenveld, this scepticism had to be overcome before gaining community support for the proposed measures.

4.4.3 Programme output: integration achieved?

The Dutch Parliament has required the Programme Directorate to report the progress and the output of the programme every six months. The 19th progress report stated on 31 December 2011, that the total cost estimate for the programme was 2170.9 million Euro compared with a budget of 2180.8 million Euro (with a margin of 10%; PDR, 2011b). This means that the expenditure until completion of the programme is expected to be between Euro 2.0-2.4 billion (price index 2011) and within the initial cost estimates (2.2 billion Euro and a bandwidth of 37%; price index 2005). Also, the 19th progress report states that by 31 December 2011, the investment decisions for 73% of programme budget had been made. Furthermore, it reports that out of the 39 initial projects that were described in the policy decision (PKB) in 2006, 5 had been cancelled because other projects will deliver greater water level reductions than expected, 8 are expected to be completed before 2015, 18 to be completed in 2015, and 8 are expected to have a delay of approximately one year (completion originally scheduled for 2015). Hence, it may be concluded that Room for the River is on track to achieve its' hydraulic targets without budget over-run or major time delay. Hence the Room for the River programme is performing, to date, significantly better than other large water programmes in the Netherlands, such as HWBP and HWBP-2 (Taskforce HWBP, 2010), and other large infrastructure projects such as the Betuweroute and the High Speed Line (Hertogh and Westerveld, 2010). In comparison, international comparative research showed that out of 258 large infrastructure projects, some 90% had cost overruns averaging some 20.4% (roads), 33.8% (tunnels and bridges) and 44.7% (rail; Flyvbjerg, 2007).

From an evaluation of the design process, it is apparent that Room for the River also meets its second programme objective of contributing to the spatial quality of the project locations (Hulsker et al., 2011). The evaluation concluded that spatial quality was successfully integrated into the water safety projects in terms of dealing with agriculture, recreation, cultural-historic values and existing residences. In some cases, the projects have provided and/or improved opportunities for urban development through better connection of both sides of the river (e.g. in Lent and Deventer). According to a large number of interviewees in my research, the rationale behind the dual objective was, besides contributing to the spatial quality of the project locations, to create local support for the measures by providing local and regional actors an incentive (improved spatial quality) for collaboration. The results of the survey indicate that this has worked well: overall, the actors involved are satisfied with results of the programme, with 85% of the respondents indicating that they were satisfied or very satisfied (total average

3.96/5; standard deviation 0.69). Similarly, a survey that was carried out for the mid-term review of the programme shows significant overall satisfaction with the results in terms of technical/design aspects of the programme (3.91/5; standard deviation 0.67; see van Twist et al., 2011a). It could, therefore, be concluded that Room for the River has an output in which water safety and spatial quality are integrated to an extent that is satisfactory to the majority of the stakeholders involved.

Both the interview and the survey data suggest that the programme management was instrumental to the delivery of the programme's output and outcome, but that the success of Room for the River cannot be attributed to the programme management alone. In the survey, respondents were asked to indicate to what extent several different factors contributed to the realisation of the programme. From most to least (average) rated importance (5 = very important, 1 = very unimportant): sense of urgency after the near floods of 1993 and 1995 (average of 4.3 out of 5); human factors, such as leadership, trust, political decisiveness (average of 4.19 out of 5); the connection of the water safety and spatial quality objectives (average of 4.03 out of 5); transparency and milestone management (average of 3.99 out of 5); centralised-decentralised set-up of organisation (average of 3.89 out of 5); contextual factors, such as previous large infrastructure projects High Speed Railway Line and the Betuwe Railway project, the economic crisis, and reorganisations of waterboards and Rijkswaterstaat (average of 3.19 out of 5). The great majority of the interviewees confirmed the importance of the governance approach of Room for the River for delivering its results. However, several interviewees also highlighted that several factors outside the influence of programme management have contributed, such as the quality of the staff involved, leadership of individuals and the economic crisis.

As most designs have not yet been implemented it is too early to draw final conclusions about the output of the programme. Nevertheless it seems that the programme is thus far successful in achieving all of its' objectives. Whilst some outcomes can be attributed as a result of Room for the River, others are the result of processes of change that were already ongoing prior to Room for the River. For example, since the 1980s, Rijkswaterstaat has increasingly emphasised the need to improve its operational excellence to make its operations more effective and efficient (see also van den Brink,

2009). However, the survey results suggest that Room for the River has contributed significantly to more intensive collaboration between different government agencies (average 3.84/5, with 1 being a very low contribution and 5 very high contribution). In addition, the interviewees shared a common view that Room for the River allocated more planning and design responsibilities to waterboards, provinces and municipalities, whilst Rijkswaterstaat took up a new role that primarily involved monitoring and facilitating to enable appropriate progress and ensure the quality of plans and designs. Overall, I have observed that the four structural factors of Room for the River's governance arrangements (Section 4.3.1.1 – 4.3.1.4) are set up to promote integrated outcomes in terms of objectives (i.e. to increase water safety whilst contributing to spatial quality). A changed economic and political context has emphasised the importance of integration to achieve increased water safety in a cost-effective and transparent manner.

At the level of spatial scales, I have also identified a coherent approach within the programme. From a hydraulic perspective, all measures are connected. After the programme office concluded that at several locations more water level reduction would be achieved than planned, it advised the Vice-Minister to cancel several projects as superfluous. Also, in individual projects, such as the Veessen-Wapenveld bypass project, solutions were sought outside the project area to overcome hurdles in the planning process (agricultural land in flood prone area was traded for a nature area outside the project area). However, this occurred only rarely. Also, with regard to spatial quality, the coherence of the measures at a river branch level could be improved (Hulsker et al., 2011).

With regards to integration across temporal dimensions, I note a difference for within and beyond the duration of the programme. Overlap between subsequent planning stages was deliberately created for early involvement of actors that are normally involved during later stages of a planning process (e.g. regulators, operation and maintenance). Involved actors suggested that this reduced the length of the planning process and improved the quality of the plans and designs. However, the national Government decided that Room for the River's water safety objective was fixed and left little room for considering higher river discharges in the future. It was decided by the national Government in the PKB that the objective for Room for the River was to achieve a discharge capacity of 16.000m^3/s at Lobith. It should be noted, however, that during the initiation phase of the programme, an evaluation was carried out to assess if Room for the River could accommodate the passage of 18.000m^3/s in the river systems in the future, so that the realised measures could retain their functionality and have a 'no-regret' performance for a discharge of 18.000m^3/s. In addition, for three measures in the river IJssel the regional stakeholders have requested an additional analysis to assess whether the realisation of the Room for the River measures can be combined with measures that already anticipate coping with a potential future discharge capacity of 18.000m^3/s (PDR, 2011a). However, further work on the potential future increase of discharge capacity is being done in the Delta Programme Rivers and is beyond the scope of Room for the River.

4.5 Discussion

4.5.1 Temporary change or transition?

In addition to being a trend breaker with regards to flood risk management, Room for the River is considered an example for the implementation of multi-level governance approaches in the Dutch and international water sector. During the commencement of the planning stage of Room for the River in 2006, it was argued by policy scientists that Dutch water management was undergoing more than just temporary change (van der Brugge et al., 2005; Wiering and Arts, 2006; Wolsink, 2006). However, it was at that time also believed that it was too early to conclude that a transition to integrated water management was complete, because there was a considerable gap between strategic policy visions for integrated water management and practical implementation which was mainly attributed to governance pitfalls related to centralised planning cultures. At present, the vision for integrated river basin management is documented in the PKB (see Section 4.3.1.2) and executed in the planning phase, which is completed for most of the measures in Room for the River. Hence, the gap between the vision and practical implementation has largely been closed. Because of the prominent role of the regional governments in the planning phase, it can also be argued that many of the governance pitfalls related to centralised planning cultures are overcome; regional governments rather than the national government took the lead in making planning decisions, whilst Rijkswaterstaat's main role in

Room for the River has been to monitor progress and quality of the plans and facilitate if necessary. This has provided opportunities to link local ambitions with the river widening projects, whilst making use of local knowledge and relationships with the community. Furthermore, the survey results suggest that these changed relationships between governments at multiple levels are likely to be permanent. However, does this mean that a transition to integrated river basin management has been completed?

The Delta Programme that was established in 2009 and is currently in its initiation phase uses Room for the River as an example (see Deltacommissaris, 2011). For example, the Delta Programme has included river widening in its portfolio of alternative options to establish long term safety against floods. Similar to Room for the River, the Delta Programme aims to integrate multiple objectives across multiple spatial scales. With regard to integration across temporal dimensions, the Delta Programme goes further than Room for the River, in the sense that it explicitly adopts the concept of adaptive management to establish effective flood risk management over the immediate and longer term (see Deltacommissaris, 2011). Furthermore, the Delta Programme has adopted a multi-level governance approach to make flood risk management a joint effort between local, regional and national government agencies and the private sector. Hence, it could be argued that Room for the River has resulted in more than a temporary change of flood risk management practice in the Netherlands.

However, based on the interview responses, I have identified a risk of losing knowledge after completion of Room for the River. According to the interviewees, much of the practical knowledge (e.g. technical knowledge about location characteristics or process knowledge about stakeholder interests, stakeholder relationships) and group dynamics that were needed to successfully complete the planning stage is tacit knowledge that is poorly documented. As such, it depends on the individuals' and teams who have contributed to the realisation of the plans. With most of those involved employed on temporary contracts for the duration of particular stages of the project (e.g. planning or realisation stage), the Room for the River approach is not (yet) firmly embedded in the working ethos of the organisations involved. It is therefore too early to conclude that a transition to integrated river basin management has been completed.

4.5.2 Is a programme an effective instrument for governing transitions?

According to the incumbents of Room for the River, it was not an official programme objective to govern the process of change. However, in hindsight, I can conclude that the programme plays an important role in the transition to integrated river basin management as I have described above. From a transition management perspective this provokes the question whether programmes such as Room for the River are effective instruments for governing transitions. Transition governance requires a mix of centralised and decentralised governance approaches (e.g. Huntjens et al., 2012) and relies on a mix of formal rules and procedures and informal interactions between individuals (e.g. Olsson et al., 2006). Based on my research findings, I conclude that the Room for the River programme entails such a mix through its arrangements for 'controlled trust' (Section 4.3.1.2 and 4.3.1.5) and facilitation to assist decentralised project teams to achieve their objectives (Section 4.3.1.3). The effectiveness of the governance configuration to govern a transition alters during different stages of a transformation process (Rijke et al., 2013). Hence, the success of a programme to govern (a part of) a transition depends on capacity to signal and anticipate changed circumstances. Room for the River's cycle of justification, monitoring and facilitation (Section 4.3.1.4) enables the programme office to do this. Furthermore, capacity building efforts in the programme (Section 4.3.1.6) play an important role in increasing the interaction between stakeholders and individuals enabling discussion of experiences, problems and solutions. Hence, I can conclude that a programme such as Room for the River can be effective for governing transitions, because it combines centralised, decentralised, formal and informal aspects and is able to shift between these according to the needs. However, as described in Section 4.4.1, there remains a risk of losing lessons learnt when the programme is complete.

4.6 Conclusion

Room for the River is successfully achieving all its objectives. It is possible to conclude that it is resulting in integrated outcomes that increase water safety whilst contributing to spatial quality. It also applies a coherent approach to spatial scales. However, the ability to successfully adapt to potential larger river flows in the future has played only a marginal role in the programme. Overall, I have observed that the four structural factors of Room for the River's governance arrangements (Vision, policy framework, economic justification, regulation and compliance; Section 4.3.1.1 – 4.3.1.4) are set up to promote and are successfully delivering integrated outcomes in terms of objectives. The four process factors (leadership, capacity building and demonstration, public engagement, research; Section 4.3.1.5 – 4.3.1.8) are enabling an integrated approach though collaborative leadership and stimulating multi-level governance approaches which are required for integrated water management.

I conclude that Room for the River plays an important role in a transition to integrated river basin management in the Netherlands. With the completion of most of the planning stage, it can be concluded that the programme has overcome the gap between strategic policy vision and practical implementation of integrated river basin management. Also, through application of a mixed centralised-decentralised governance approach, the programme has tackled governance pitfalls related to centralised planning approaches that previously impeded integrated water management. I argue that a governance approach as applied in Room for the River can be effective for governing transitions, as it combines centralised, decentralised, formal and informal aspects and is able to shift between these according to the needs. However, I also have identified a risk of losing many of the lessons learnt when the programme is complete. Hence, it could be concluded that, since the commencement of the planning stage of Room for the River in 2006, the main challenge in terms of transition management has shifted from removing impediments to establishing continuity of the newly introduced governance approach.

CHAPTER FIVE

Adaptive programme management through a balanced performance/strategy oriented focus.

Delivering adaptation projects.

This chapter addresses the 'how' question by focusing on how large scale adaptation projects can be delivered effectively. A set of attributes for the effective delivery of large scale adaptation projects is presented in this chapter. These attributes are specifically developed for policy makers, programme managers and project managers who are involved in setting up and managing of large flood protection programmes.

5. Adaptive programme management through a balanced performance/strategy oriented focus

This chapter is adapted from:

Rijke, J., van Herk, S., Zevenbergen, C., Ashley, R., Hertogh, M., ten Heuvelhof, E. (in press) Adaptive programme management through a balanced performance/strategy oriented focus. *International Journal of Project Management*.

Abstract

This chapter explores how programme management (as opposed to project management) can contribute to the effective design and delivery of megaprojects. Traditionally, project management is considered to be performance focused and task oriented, whilst programme management entails a more strategic focus. The programme management literature suggests that this can results in tensions between the management of the projects and the programme as a whole. This paper uses the findings of the €2.4 billion Room for the River flood protection programme in the Netherlands as a case study, because indicators about its budget, time, quality and stakeholder satisfaction suggest high programme management performance upon completion of the planning and design stage of its 39 river widening projects. Based on a literature review, document analysis and 55 face-to-face interviews, I have analysed how the programme management of the programme contributed to this result. Six attributes for effective programme management that are identified from the project and programme management literature are used to structure the research data. Consecutively, the interactions between project and programme management are analysed. The analysis of Room for the River reveals a combined strategic/performance focus at the level of both programme and project management that enables a collaborative approach between programme and project management. This particularly enables effective stakeholder collaboration, coordination and adaptation of the programme to contextual changes, newly acquired insights and the changing needs of consecutive planning stages, which positively contributes to the performance of the programme as a whole.

5.1 Introduction

Megaprojects in the infrastructure sector often fail to meet project objectives within their initial budget and time constraints. For example, international comparative research showed that out of 258 large infrastructure projects, some 90% had cost overruns averaging between 20% and 44% for different types of projects (Flyvbjerg, 2007). In addition, the European Commission repeatedly reported time delays in realising the programme of the Trans-European Network, that consists of 30 priority projects with a total investment of € 600 billion (estimation 2005; Hertogh and Westerveld, 2010). The performance of megaprojects depends partly on their management (Hertogh and Westerveld, 2010; Kwak et al., 2013; Shao et al., 2012). In addition, changing context factors during the long planning period of these projects can lead to delays and cost overruns when they complicate the accomplishment of important project decisions and/or change the scope of large infrastructure projects (Hertogh et al., 2008; Pellegrinelli et al., 2007). Furthermore, inadequate balance between project control and stakeholder engagement could lead to illegitimate project decisions that lack stakeholder support or create false expectations, both potentially resulting in cost overruns, inadequate progress and poor quality (Hertogh and Westerveld, 2010). Especially when tensions between stakeholders occur, project organisations tend to lean on the control approach, which often leads to disappointing results (Hertogh and Westerveld, 2010).

Megaprojects, particularly in the infrastructure sector, are often being managed as programmes, because megaprojects typically consist of multiple components that can be classified as sub-projects (Pellegrinelli, 1997). A programme can be defined as *"a group of related projects managed in a coordinated way to obtain benefits and control not available from managing them individually"* (Project Management Institute, 2008a, p.434). Although it is in practice difficult to make a clear distinction between a programme and a project, programme management is more than a scaled-up version of project management, because programmes include elements that are outside the scope of individual projects within a programme (Lycett et al., 2004; Maylor et al., 2006). With this, programmes are increasingly being adopted to implement organisational transformational strategies and integrate multiple projects (Maylor et al., 2006). Furthermore, the roles of programme managers are more strategic by nature compared to those of project managers who are more task oriented and performance driven (Brown, 2008a). This difference often creates tensions between programme and project management and can hinder the achievement of project and programme objectives (Lycett et al., 2004).

This paper contributes to the programme management literature by exploring how project and programme management can collaborate effectively. The 2.4 billion Euro flood protection programme Room for the River in the Netherlands is used as an illustration, as this programme has performed relatively well during the initiation and planning/design stages in terms of output, stakeholder satisfaction, budget and time (Rijke et al., 2012c). For example, many other recent flood protection programmes were significantly more expensive than anticipated (e.g. Kim and Choi, 2013; Taskforce HWBP, 2010). The lessons that are presented in this paper potentially carry broad international relevance, because many large scale infrastructure upgrades to protect against flooding from the sea and rivers are organised through programmes that include multiple projects (Zevenbergen et al., 2012). As there are many similarities of flood protection programmes with large infrastructure projects in other sectors (Hertogh et al., 2008), the relevance of this paper could potentially also stretch beyond the field of flood management.

5.2 Theory

5.2.1 Project versus programme management

There is an emerging body of literature about programme management that originates from the project management literature, but has several theoretical bases such as organisational theories, strategy, product development manufacturing and change (Artto et al., 2009). As such, there are many different interpretations to the meaning of programme management (Artto et al., 2009; Pellegrinelli et al., 2007). The traditional view of programme management is an extension of project management and focuses primarily on the definition, planning and execution of a specific objective (Lycett et al., 2004; Pellegrinelli, 2002, 2011; Pellegrinelli et al., 2007). In this view, programme management is a mechanism to coordinate the performance of a group of related projects (Ferns, 1991; Gray, 1997). A more recently developed view stems from strategic planning and attributes a broader role to programme management in terms of value creation for the organisations involved beyond the performance of projects in a particular programme (Murray-Webster and Thiry, 2000; Thiry, 2002, 2004; Young et al., 2012). Overall, programme management is used to create portfolios of projects (Gray, 1997; Turner, 2000), implement strategies (Partington, 2000; Partington et al., 2005) and generate change in products, business or ways of working (Pellegrinelli, 1997; Ribbers and Schoo, 2002; Thiry, 2004).

Programmes consist of multiple projects that run in parallel or (partly) sequential (Lycett et al., 2004; Maylor et al., 2006). However, the relationship between a programme and a project differs from the relationship between a project and a work package, as programmes can provide benefits over and above those that projects can achieve on their own, such as improved exposure, prioritisation, more efficient use of resources and better alignment with other projects (Pellegrinelli, 1997). Whilst project management is typically focused on performance in terms of quality, cost and time, programme management operates more on a strategic level to create synergies between projects and deliver a package of benefits through coordination of a series of interconnected projects (Lycett et al., 2004; Maylor et al., 2006). As such, programme management requires a different approach than project management (Partington et al., 2005), that takes a broader organisational scope and takes into account the interactions between projects (Maylor et al., 2006; Shao et al., 2012; Young et al., 2012).

Different types of programme management exist in which the programme management interacts differently with the management of individual projects (Lycett et al., 2004; Pellegrinelli et al., 2007; van Buuren et al., 2010). For example, programme management can take the form of as portfolio management, a shared service centre and goal-oriented programme management. In case of portfolio management, programme management contributes to a higher level fine-tuning of project ambitions and prevents fragmentation in decision-making, without altering the planning and budget cycles of individual projects (Gray, 1997; van Buuren et al., 2010). This typology is applied to coordinate the effective use of resources, risk management and branding of a group of multiple projects (Gray, 1997). As the project objectives are often leading in this typology, programme management has limited influence on the internal management of individual projects which mutually adapt based on open information (Gray, 1997). When acting as a service centre, programme management can coordinate the management of knowledge across multiple projects through integration of, for example, financial, legal, administrative and technical services into a 'shared service centre' (van Buuren et al., 2010). Programme management takes a more dominant role in a goal-oriented programme management style in which programme objectives prevail over the objectives of individual projects and case selection, prioritization and adjustment occurs integrally to achieve the overarching programme ambition (Ferns, 1991; Pellegrinelli, 1997). In practice, programme management often occurs as a combination of different typologies (Pellegrinelli et al., 2007).

The performance focus of project management and the more strategic focus of programme management complement each other (Pellegrinelli, 2011; Thiry, 2002). However, because of these different management perspectives, a number of issues are common place at the interface of programme and project management within programmes (Lycett et al., 2004). For example, the question is "who is in charge?" refers to what extent programme management should be limited to providing support or, by contrast, imposing objectives and ways of working on individual projects (van Buuren et al., 2010; p.680). An inappropriate degree of control by programme management is counterproductive (Hertogh and Westerveld, 2010; Lycett et al., 2004). Too much control can impose excessive bureaucracy upon project management, resulting in diversion of project management resources from achieving project objectives (van Aken, 1996; van Buuren et al., 2010). This is often combined with a programme management focus at an inappropriate level of detail, which diverts programme management resources away from their strategic function. As such, excessive control could compromise the relationship between programme and project managers, causing rigidness and diverting energy from value adding activities (Lycett et al., 2004). When followed by a response of more bureaucracy and control, this could invoke a vicious circle of bureaucracy and de-motivation and inflexibility (Platje and Seidel, 1993). However, insufficient control may, for example, cause loss of synergies between projects and therefore reduced quality, cost overruns and delays (e.g. Lycett et al., 2004; Maylor et al., 2006; Unger et al., 2012). Other common challenges for programme management are insufficient flexibility for programme management to adapt to changes in the context of individual projects (Lycett et al., 2004; Sanderson, 2012); and cooperation between projects within a programme tends to be difficult due to inter-project competition and failure to harness organisational learning (Lycett et al., 2004).

5.2.2 Attributes for effective programme management

The emergence of programme management literature has logically been accompanied with research and guidelines about managing programmes effectively (e.g. Project Management Institute, 2008b; Shao and Müller, 2011; Shao et al., 2012; Shehu and Akintoye, 2009; UK Office of Government Commerce, 2007). The success of programme management can be examined at the project, programme and corporate level (Blomquist and Müller, 2006; Jonas, 2010; Müller et al., 2008). Traditionally, programme effectiveness is measured by time, cost and performance, sometimes complemented with satisfaction of stakeholders, users and the programme team (Shao et al., 2012). It should be noted, however, that programme effectiveness is more recently also measured in terms of capability to change organisations, to increase organisations' market share and to innovate, as the focus of programme management literature shifted to a more strategic orientation (Shao et al., 2012). Consistent with this broader view on programme effectiveness, Shao and Müller (2011) have identified six dimensions for programme success: programme efficiency, impact on programme team, stakeholder satisfaction, business success, preparation for the future, and social effects. In addition, Shao and Müller (2011) have developed a set of dimensions for programme context that influence programme success, including stability of the context, harmony between a programme and its context, support for a programme, and adaptability of a programme to its context.

Several attempts have been made to identify attributes for effective programme management, resulting in attributes for programme design and attributes for programme management processes (e.g. Hu et al., 2012; Shao and Müller, 2011; Shao et al., 2012; Shehu and Akintoye, 2009). The foundation for programme success is laid during the initiation stage, when opportunities are discovered and ideas are created and transformed into a programme design (Heising, 2012). Stakeholder collaboration plays an important role during this stage to align objectives, roles and responsibilities of stakeholders involved and position and formalise the ideation strategically in a way that a broadly supported *programme vision* is developed with overarching programme goals, and a *programme priority focus* that allocates resources to these goals (Heising, 2012; Sanderson, 2012; Shehu and Akintoye, 2009). In addition, a *programme planning framework*, facilitates a shared understanding among programme stakeholders through guidance of the execution and control of the programme is needed to provide integration between the initiation stage and the consecutive stages of the programme (Heising, 2012; Reiss et al., 2006; Shehu and Akintoye, 2009).

Furthermore, the role of a programme management office (e.g. a project portfolio management organisation) is considered key to achieving programme management success, because it enables integration of resources throughout the system of projects through programme governance, coordination and adaptation (Davies and Mackenzie, 2013; Unger et al., 2012). *Programme governance* refers to the process of aligning internal programme stakeholders and anticipating external stakeholders so that the programme strategy is executed efficiently and value is added to the individual projects and programme as a whole (Beringer et al., 2013; Too and Weaver, 2013). As such, it is closely related to the programme planning framework and strategy that are developed during the initiation stage. *Programme coordination* is needed for the coordination of tasks, control of performance and support of project teams (Chen et al., 2013) and depends on information availability, goal-setting and systematic decision making in both projects and programmes (Martinsuo and Lehtonen, 2007; Teller et al., 2012). *Programme adaptation* is needed to address or anticipate contextual changes (Ritson et al., 2011; Shao and Müller, 2011) and depends on the fit between the programme and organisational strategies, the flexibility of programme structures and procedures and the adaptability of a programme to its context (Shao et al., 2012).

Whilst programme governance, coordination and adaptation are considered as different success factors, they are connected and cannot be seen in isolation of each other, because of mutual dependencies between these factors and the need for approaches that are both robust and flexible. For example, there should be a balance between control and stakeholder engagement to achieve legitimate and supported decisions and desirable programme performance (Hertogh and Westerveld, 2010). Finding a balance between robustness and flexibility of programme management approaches to deal with changing dynamics and contexts is considered one of the most challenging aspects of programme management (Davies and Mackenzie, 2013; Sanderson, 2012).

5.3 Research approach

The aim of this paper is to explore how programme management (as opposed to project management) can contribute to the effective design and delivery of megaprojects. This paper focuses on programme design and interactions between the management of individual projects and the programme as a whole.

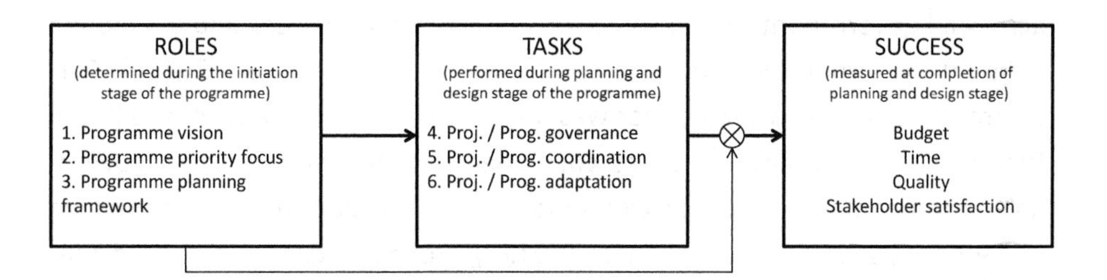

Figure 5.1 Analytical framework for measuring programme management effectiveness (adapted from Jonas, 2010)

Figure 5.1 presents the analytical framework for this paper to analyse the effectiveness of programme management process in which the programme management office and project teams interact. It is based on the assumption that programme management effectiveness is determined by the quality of fulfilling roles by the various tasks of programme management office and project teams (Jonas, 2010). I acknowledge the possibility of other causes for programme success, such as inadequate preconditions (e.g. insufficient funding), unavailability of required competences or the impact of contextual factors, but leave these outside the scope of the study as I focus on the interaction between programme management and project management within a certain programme. The analytical framework includes all attributes for effective programme management that are described in the previous section. I have used the six attributes of the framework to describe how project and programme management interacted in the case of the Room for the River programme in the Netherlands. Attributes 1-3 are determined during the initiation stage of the programme (programme design), whilst attributes 4-6 are performed during the planning and design stage of the programme by both the programme management office and the project teams.

The 2.4 billion Euro flood protection programme Room for the River in the Netherlands has been selected as a case study, because the programme has reported, upon completion of its planning and design stage, that it is on track to achieve its objectives within budget and without major delays (excluding the project IJsseldelta, which was added to the programme in 2012 and is scheduled for completion in 2019, four years behind the initial time schedule; PDR, 2013). In addition, the majority of individuals who were actively involved in the programme (e.g. decision makers and project officers across all government levels) were satisfied with the process and developed plans of the programme (Rijke et al., 2012c; van Twist et al., 2011b). Furthermore, based on a survey that was held amongst participants of the Room for the River programme, it was concluded that the programme's governance arrangements were instrumental in the programme's performance (Rijke et al., 2012c; van Herk et al., 2012d). Hence, it can be assumed that programme and project management interacted effectively during these stages. Because the most complex management processes typically take place during the initiation and planning/design phases of large scale infrastructure programmes (Hertogh and Westerveld, 2010), the Room for the River programme provides a good opportunity to study how the programme management processes have evolved and how programme and project management interacted during these phases of the programme (initiation: 2000-2006; planning/design: 2007-2012). However, conclusions about the programme's effectiveness for achieving objectives should be finally considered when the realisation of the programme is completed.

It is relevant to note that Room for the River's approach to flood risk management and governance is considered exemplary in a national and international context (Kabat et al., 2009; van den Brink, 2009; Zevenbergen et al., 2012). It is the first program in the Netherlands that breaks with a long history of reducing the space for river to flow (Rijke et al., 2012c; van der

Brugge et al., 2005). Instead, it creates at 39 locations more space for the rivers using measures such as floodplain excavation, peak discharge channels and dike relocation. The 39 projects were managed individually, whilst a programme office was set up at Rijkswaterstaat to manage the programme as a whole. With this number of individual projects, the programme inherently contains elements of both programme and project management and is thus suitable for analysing interactions between both levels of management. A combination of a goal oriented and a service centre programme management style is applied to Room for the River, as the programme objectives prevail over the objectives of the individual projects and a programme office is set up to actively assist the individual projects in achieving these objectives (Rijke et al., 2012c). Furthermore, Room for the River is the first large scale infrastructure programme in the Netherlands that has adopted a multi-level governance approach in which the traditional hierarchical governance approach is replaced by an approach that combines centralised and decentralised steering processes: the decision frameworks of the programme for establishing improved flood protection and landscape quality are set by the national government, whilst the 39 designs are prepared and most decisions taken by local and regional stakeholders (Rijke et al., 2012c; van Herk et al., 2012b).

A combination of document analysis and face-to-face interviews (n=55) have been used to analyse the governance activities within programme management, project management and the interface between programme and project management and their implications for Room for the River's performance. Interviews were conducted with individuals who were involved with the initiation (n=10), design and realisation (n=31) stage of the programme, as well as individuals in strategic positions at the levels of senior policy maker and decision maker (n=14). Interviewees represented a range of disciplines and organisations involved with the individual projects and the programme as a whole. As such, they represented waterboards, provinces, municipalities, the Ministry of Infrastructure and Environment and its executive organisation Rijkswaterstaat, including the Room for the River programme office which is staffed by Rijkswaterstaat.

All interviews were semi-structured and covered similar topics: the motivation for the design of the organisation of Room for the River, project and programme management activities, and collaboration between stakeholders in the programme and projects, and the translation of knowledge developed in Room for the River to other programmes and organisations. The interviewers made sure that all topics were covered during each interview in order to enhance comparability between the different interviewee perspectives. However, the interviewees' responses steered the flow of the conversation, allowing the researchers to make best use of the interviewees' knowledge by giving the opportunity to the interviewees to elaborate on the topics that they deemed most relevant and to avoid that any explanation for the effectiveness of the programme management was overlooked.

The analysis of the interview data comprised of two steps. Firstly, the data were structured according to the six attributes for effective programme management (Figure 5.1). Secondly, a series of validation sessions were organised - including a workshop with officials of various government agencies (28 participants) and a workshop with a user panel comprising senior policy advisors (5 participants) – in which the participants were asked to reflect upon the attributes for effective programme management. During these sessions, the participants discussed and confirmed the results, emphasising the importance of programme governance (stakeholder collaboration) and programme adaptation in particular. In addition, observations that were made at three training sessions about stakeholder management (45 participants), two political conferences (approx. 220 participants) and a network event for the diffusion of lessons learnt from Room for the River within Rijkswaterstaat (approx. 150 participants) confirmed the results.

5.4 Case Room for the River

In this Section, the six attributes for effective programme management (Figure 1) are used to describe the management of the Room for the River programme during the initiation stage and the consecutive planning and design stage.

5.4.1 Programme vision – Flood risk reduction through river widening

After near-miss river floods in 1993 and 1995, which in 1995 led to the evacuation of 250,000 people and 1 million cattle in the Netherlands, the awareness increased amongst the public, politicians, public administration and water professionals that nature cannot be controlled by the traditional way of canalising and that new ways of managing rivers was required; i.e. through creating more space for rivers to discharge their flows. However, this vision of making more space for rivers was on several occasions questioned by engineers, economists and politicians. For example, opposing engineers argued that the traditional approach of dike improvement has been successful for a long time and that river widening was not a proven method to provide better or cheaper solutions (e.g. Vrijling, 2008). Further criticism was that a cost-benefit assessment of the approximately 700 measures that were initially considered in Room for the River's initiation phase suggested that the programme was necessary and beneficial, but that creating more space for rivers was not the cheapest option for all river branches (Ebregt et al., 2005; Eijgenraam, 2005).

However, interviewees who were involved with the initiation stage of the programme commented that, at the time of initiation of the Room for the River programme, political decision makers were not convinced by these criticisms and decided to go ahead with the programme because they argued that river widening would add more value to the river area (e.g. economic, nature, recreation) and was a more effective flood risk measure (as failure of higher dikes would result in more water in polders and thus more damage). In addition, stakeholder engagement played an important role in translating the dominant political and policy perceptions to a legitimate programme vision. For example, a politician who recalled broad societal resistance against dike strengthening in the 1970s commented: *"dike strengthening has a much higher societal impact than river widening."* In addition, an executive at Rijkswaterstaat commented: *"The basic idea was supported by a broad political movement. This was needed for being successful."* Hence, the vision for flood risk reduction through river widening was reflected in the Room for the River Policy Decision (PKB Ruimte voor de Rivier), which came upon the completion of the initiation stage of the programme (see also Section 4.3). This PKB included the programme's formal (dual) objectives of: 1) improving safety against flooding of riverine areas of the Rivers Rhine, Meuse, Waal, IJssel and Lek by accommodating a discharge capacity of 16.000m3/s; 2) contributing to the improvement of the spatial quality of the riverine area.

Overall, the interviewees held a common view that the new vision and dual objective of the programme were instrumental to the programme success, because they provided something to gain in the individual projects for all stakeholder groups involved.

5.4.2 Priority focus – Broad stakeholder support through connecting decentralised ambitions

The central programme management office (PDR) that was established after the completion of the PKB in 2006 aimed for satisfactory delivery of programme objectives. During the initiation stage, it was decided by the individuals who were involved with the programme design of Room for the River that a multi-level governance approach was needed to achieve this. During the interviews, they expressed two main motivations for this choice: 1) they were aware that river widening required close collaboration between water management, spatial planning, and other disciplines such as ecology and landscape architecture; 2) top-down governance approaches in delivery of the railway projects Betuweroute and the High Speed Line provoked resistance amongst local communities and politicians and led to delays which created widespread community scepticism about large infrastructure projects. Instead, the 'architects' of the Room for the River programme argued that close collaboration between governments at various levels was needed. Formalising spatial quality as a formal programme objective, created incentives for local/regional governments to engage with the PDR and strive for synergetic outcomes in terms of agriculture, urban development, recreation, nature and cultural-historic values.

As a result, decentralised government agencies (i.e. provinces, waterboards and municipalities) took up a lead role in managing the majority of the projects in the planning phase to ensure that their interests were well represented in the planning and design processes. For example, a local politician whose municipality was responsible for a project during the planning phase explained: *"We wanted to be in the lead of the project to stay in control during the design process and protect the interests of the companies in our area"*. Also, a decision maker from a Province commented that *"because of the dual objective in the PKB, it was easier to involve the Province"*, whilst a staff member of the programme office suggested: *"The spatial quality objective was useful process input. The integration in the design depended on the initiator of the project, who could fit in the water safety works with other functions in the project area"*. It should be noted, however, that the dominant view amongst the interviewees was that improved flood protection rather than spatial quality was the leading driver for the projects and remained so during the development process. It can, therefore, be concluded that the dual objective created wide support for the proposed projects.

With this, the programme's priority focus provided a clear and legitimate basis for effective stakeholder collaboration within the individual projects throughout the duration of the programme.

5.4.3 Programme planning – Strict boundary conditions and clear roles

In line with the advice of the Commission Elverding, which urged large infrastructure projects in the Netherlands in 2008 to apply improved planning processes for *"faster and better"* results (Commissie Elverding, 2008), Room for the River aimed to deliver the proposed measures before 2015 through stable project decisions throughout the project. To avoid delays, the vision was to involve politicians and non-governmental stakeholders early in the planning process to establish commitment and support, and to deliberately create overlap between separate planning stages (initiation, planning, realisation) to generate input early in the planning process from actors responsible for regulation, operation and maintenance in order to establish realistic plans and designs of good quality. For example, a politician that was involved during the initiation explained: *"It's better to let the crowd think along during the early phases of a project, because this saves a lot of trouble in later phases, provides new insights, makes the bottlenecks immediately clear, and makes it morally difficult to go into appeal if you have been involved."* The transitions between planning phases were supported through agreements between the key stakeholder groups involved with performing the most important activities during the following planning phase to enhance the stability of the project decisions and secure collective leadership. For example, a financial controller at the programme office explained: *"Thinking in advance about future planning phases is important to be able to facilitate projects effectively. For example, during the design phase, we help the regional initiators to understand the value of timely making strategies for tendering and procurement of the construction works."*

The Room for the River Policy Decision (PKB Ruimte voor de Rivier) was produced upon completion of Room for the River's initiation phase (2000-2006). This document was established after extensive consultation with key stakeholders (see also ten Heuvelhof et al., 2007) and outlined the vision, programme objectives (see section 5.4.1), a selection of 39 locations for river widening, the types of measure that needed to be implemented at these locations, as well as the procedures for the planning and realisation phases of the programme and the roles and responsibilities of the stakeholders. It described the principle that decentralised steering and execution of tasks should be applied where possible. Nonetheless, a central programme office (PDR) was established at Rijkswaterstaat to monitor progress, quality of plans, and achievement of objectives. With this, the PKB document described a goal oriented programme management style for the Room for the River programme that would be based on a steering philosophy of *"controlled trust"*. The interviewees held a common view that the PKB remained important during the design stage of the programme, because it provided a point of reference by documenting the vision, objectives, procedures, roles and responsibilities in a document that was supported and co-signed by all the governments involved.

Formal arrangements played similar important roles during the transitions to later stages of the programme. For example, political covenants for initiation of projects (political agreement), collaboration agreements for the planning phases (cooperation agreement) and realisation agreements for the realisation phases (realisation agreement) define the collaboration by setting out the ambitions and responsibilities. In all projects within Room for the River, the national government is the client. In the initiation stage, the programme office executes the assignment. During the planning phase, each project was assigned an 'initiator' (i.e. province, waterboard or municipality) who had formal responsibility to complete the plans. During the realisation phase, a 'realisator' (preferably the organisation responsible for maintenance of the new plan) has been responsible for acquiring permissions, tendering and contracting private parties for the implementation of the plans.

Summarising the above, it can be concluded that the programme planning positively affected the effectiveness of both the individual projects and the programme as a whole by creating transparency about the roles and responsibilities during the consecutive stages of each of the individual projects in the programme.

5.4.4 Programme governance - Decentralised decision making within centralised boundaries

As described in sub-sections 5.4.1-5.4.3, programme governance played an important role to involve actors who would become internal programme stakeholders (i.e. organisations that had a direct role in project management teams or in the management and control of the programme as a whole) and external stakeholders (e.g. NGOs, community groups, lobby groups, private sector) after the completion of the initiation stage.

The aggregate interview data suggest that the mixed centralised-decentralised governance approach that was formulated in the PKB was indeed taking place during the planning and realisation phase of the programme (Table 5.1). Centralised activities in Room for the River included primarily monitoring the progress and quality of the 39 individual projects (i.e. monitoring of budget, time, project risks, hydraulic performance, spatial quality, soil management, legal issues, coherence of design). Later, after the PDR identified that time and resources could be used more efficiently, it also has been facilitating the projects by providing guidelines for issues that are common across multiple projects (guidelines are, for example, prepared for topics for spatial quality, underground cables and pipes and risk assessment) and (ad hoc) expert knowledge on all aspects that are monitored. Also for individual projects, the PDR assists project teams discussed with decision makers at national, regional and local levels and brought together various stakeholders. If the programme-wide delivery of projects was inhibited by existing policy and legislation, the programme office discussed this with national policy makers and legislators. Furthermore, the PDR enhanced the capacity of the project teams through training sessions (e.g. risk management, process management) and network events (e.g. political conference, project leaders day, stakeholder managers day). All other aspects of the planning and realisation of the measures have relied on the decentralised management of the projects. As such, problems and potential solutions have been explored including with local/regional stakeholders. This has resulted in collaborative learning processes that have, in most cases, created mutual trust amongst stakeholders and led to broadly supported designs (see also van Herk et al., 2012b).

Table 5.1 Key management activities in the Room for the River programme

Centralised activities *Key activities performed by central programme office (PDR)*	**Decentralised activities** *Key activities performed by regional project teams*
Funding of the planning and realisation of the water safety measures Monitoring and quality control Standardization of project management Facilitation/knowledge management Influencing national policy and legislation Capacity and network building amongst professionals involved Gaining political support of national stakeholder groups In case of conflict, bringing local/regional stakeholders together Justification of progress to Ministries and RWS Communication about programme	Co-funding for add-ons to the water safety measures for secondary purposes such as recreation, nature and tourism. Planning, design and engineering Justification of progress and decisions to national government through PDR Procurement and tendering Community engagement Gaining political support of regional stakeholder groups Communication about projects

The distribution of activities among the partners involved, as described in Table 5.1, was negotiable: if local partners (provinces, water boards, municipalities) claimed a stronger position than initially agreed in the PKB (more tasks, more responsibilities), this was open for discussion. This has resulted in a tailor made implementation structure, which reflects the relationships amongst the actors involved and which differs across the various projects. For example, a mayor whose city took the lead in the planning process of a measure in his area said: *"We wanted to be in the lead of the project to stay in control during the design process and protect the interests of the companies in our area."*

In summary, it can be concluded that the programme strategy of decentralised decision making within centralised boundaries contributed to the programme effectiveness by providing project managers the opportunity to align projects to their local context as long as the overall programme objectives would be achieved. This enhanced the legitimacy of both programme and projects, which in turn increased the effectiveness of both project and programme management.

5.4.5 Programme coordination – A cross-disciplinary approach to milestone management

A balanced 'triangle' of project management, stakeholder management and technical knowledge management is being applied throughout the whole programme. The PDR contains three major departments for project control & risk management, process management and technical knowledge management (as well as a communication department). Respondents working for PDR suggested that balancing between performance (i.e. budget, time, risk), quality (i.e. hydraulics and spatial quality) and legitimacy and stakeholder commitment was a key success factor for the effectiveness of the programme. They argued that maintaining such a balance prevented overemphasising one of the three aspects which could lead to unrealistic planning schedules, inadequate empathy with regional projects or unsatisfactory quality of project outputs (if project management was too dominant), unrealistic ambitions and expectations of regional stakeholders and project delays (if stakeholder management was too dominant), and too intensive involvement of PDR with regional projects (if technical knowledge management was too dominant).

Similarly, the regional project teams have been structured according to Rijkswaterstaat's standardized 'Infrastructure Project Management' (IPM) roles and comprise an overall project manager, technical manager (knowledge management), stakeholder manager (process management with regional stakeholders), contract manager (procurement, tendering and contracting) and project controller (project performance and risk). Because the regional governments that were involved with the management of the projects were not familiar with the IPM model, interviewees from these organisations indicated that they felt that Rijkswaterstaat had forced the IPM model on them. However, they also indicated that it did not differ much from what they called *"normal"* configurations of project teams. Furthermore, interviewees from the project teams and the PDR indicated that having counterparts in project teams and PDR enabled effective cooperation and that it enabled the formation of informal networks of individuals that fulfil similar roles in different projects. This enhanced the knowledge sharing amongst the individuals across various organisations within and between the various levels of government.

Furthermore, the interviewees commonly suggested that the steering philosophy of *"controlled trust"* enabled a collaborative approach amongst project and programme managers. At the core of this approach has been a proactive 'justification cycle' that enabled the PDR to monitor progress and quality of the projects, facilitate projects where needed, and justify projects to Parliament. A milestone management procedure is used for the monitoring and facilitation processes in this cycle. According to this procedure, the project teams need to follow a pre-set design process with intermediate milestones for which products had to be delivered by the project teams to the PDR (e.g. alternative design options, preferred design, final design and supplementary material) to the program office accordingly. It is the task of PDR to evaluate the quality of these products in terms of hydraulic performance, spatial quality, legal procedures, soil, integrated design, budgets and risk management. Every 6 months, the programme office has been required to send a progress report about the programme as a whole to the Dutch Parliament. Because 'river branch managers' were assigned by the programme office to have regular interaction with project teams, this management arrangement was instrumental in pro-actively signalling (potential) problems and taking remedial action (Section 5.4.6). However, several project leaders who were interviewed indicated that the requirement to report progress is resource intensive for the project teams.

Summarising the above, it can be concluded that a balanced coordination of the programme supported the achievement of realistic, feasible and broadly supported project output and enabled learning throughout the whole programme.

5.4.6 Programme adaptation – Continuous alignment with programme and project contexts

The case study finding suggest that programme adaptation was the results of robustness and flexibility that were incorporated in the initial programme design and a combination of programme governance (section 5.4.4) and coordination (section 5.4.5) throughout the programme.

Many interviewees in programme and project management positions suggested that intensive stakeholder collaboration during the early stages of the programme enabled effective adaptation during later stages as this provided a robust basis and flexibility to adapt. In particular, the legitimacy of the programme vision and priority focus and clear planning framework of the programme design supported a stable basis (i.e. robustness) for the programme management processes. The programme design provided the flexibility to adapt to these changes through the terminology that was used in the programme vision and priority focus (particularly the spatial planning objective; section 5.4.1-5.4.2) and the planning framework (section 5.4.3). For example, the terminology of the dual objective enabled adaptation of the plans to the changed political landscape after the national elections in 2010 that demanded more cost-effective measures and minimising the loss of agricultural land by highlighting its flood protection objective and changing its interpretation of spatial quality. In addition, the combination of the planning framework, in which the programme objectives prevail over the project objectives, and programme wide monitoring of achievements enabled cancelling of projects which became redundant after other projects realised more water level reduction than initially planned.

Moreover, the allocation of the responsibility for the delivery of the designs to local and regional government levels (sections 5.4.3-5.4.4) meant that these government levels became responsible for customising solutions to their own context and dealing with local stakeholders. Because these organisations are typically more knowledgeable and better connected with the regional context than the programme management office which was operating on a national level, this enhanced the ability of the programme to adapt to the context of individual projects. In addition, frequent monitoring of projects' achievements through the milestone management procedure (section 5.4.4) contributed to signalling emerging issues and addressing these through centrally coordinated learning and capacity building activities. A staff member of the PDR illustrated this: *"Around 2006, many projects scored insufficiently on the topics of permits, cost estimates and soil. We addressed this by attempting to establish pro-active risk management and quality improvement of designs through facilitation... ....which, for example, occurred through clarification of issues, seeking and/or providing expert advice, sharing experiences and instructing project teams, and providing design manuals and customised training programmes."* Adaptation therefore improved the efficiency of the management of the individual projects and the programme as a whole.

The interview data suggest that another important factor for learning and programme adaptation were audits and evaluations conducted by independent experts. Independent auditing and evaluation of performance, processes and quality of output have played an important role in the programme to reach consensus about important decisions during the planning process and generate supported programme output (Klijn et al., 2013). For example, the programme management commissioned independent organisations and commissions to audit the hydraulic performance and spatial quality of the proposed river widening designs and to advise about risk management throughout the whole programme. Furthermore, the PDR used evaluations carried out by policy scientists and consultants regarding design processes (Hulsker et al., 2011) and the decision making processes and outcomes of Room for the River (ten Heuvelhof et al., 2007; van Twist et al., 2011b) to confirm decisions and to adjust the governance arrangements where necessary.

It should be noted that adaptation of the programme management to changing contexts safeguarded the effectiveness of programme governance under changing circumstances. Lessons from individual projects were scaled up to the programme as a whole through the development of guidelines (e.g. for soil, cables and pipes, spatial quality) and coordinated network building activities amongst project teams and individuals with similar roles throughout the programme (e.g. project leaders, controllers, process managers, communication staff). Interviewees suggested that this created an incentive for the programme office to establish a community of Room for the River professionals in different organisations involved that strengthened stakeholder collaboration throughout the whole programme and stimulated learning between projects.

Summarising the above, it can be concluded that the combination of programme governance and coordination enabled programme adaptation, which has resulted in continuous alignment of the management of programme to the context of the programme as a whole and the contexts of the individual projects.

5.5 Discussion

5.5.1 Collaboration between programme and project management

The research findings show that the programme office and project teams both had a combined focus on strategy and performance. The vision, priority focus and planning framework were developed in a way that matched with the programme's overall context. Similarly, the plans and designs of the 39 projects were developed by the project teams in collaboration with the stakeholders of their local context. As described in sections 5.4.4 and 5.4.6, stakeholder collaboration played a critical role in establishing legitimacy, spatial quality and effective delivery of the programme's plans and designs. If the context demanded adjustment of the programme's organisation or measures, the programme office would coordinate programme wide adaptation. The coordination of the performance the individual projects and the programme as a whole through justification, monitoring and facilitation of progress and quality is at the heart of the performance focus of both the programme office and the project teams. Figure 5.2 summarises the relationships between the programme office, project teams and national and local/regional contexts.

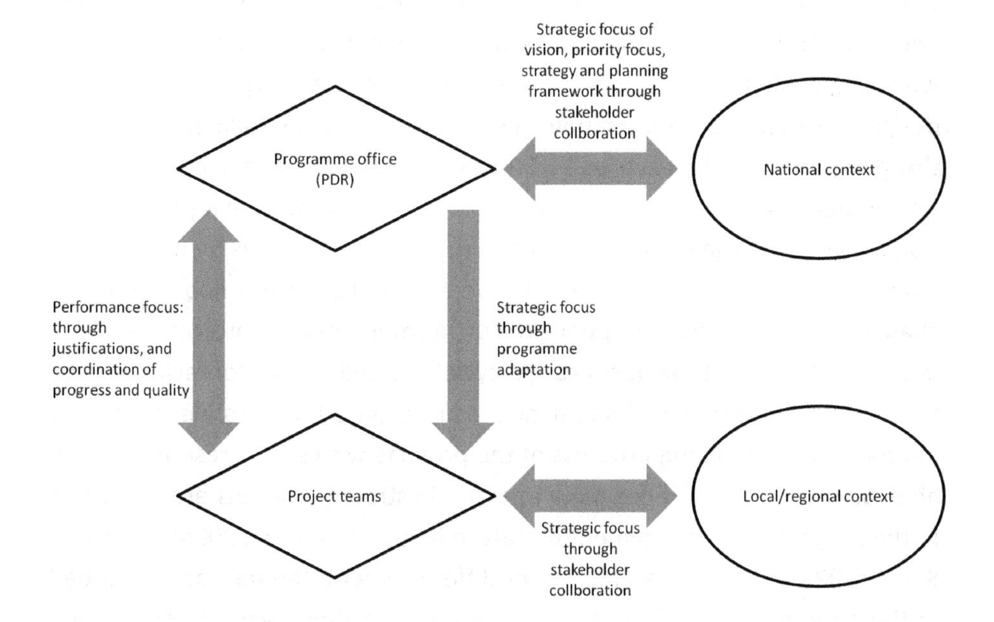

Figure 5.2 Alignment of performance and strategic foci within in Room for the River

It can be concluded that the management of the individual projects and the programme as a whole indeed both held strategic and performance foci. As such, Room for the River contains elements of the traditional (e.g. Ferns, 1991; Gray, 1997) and more recent (e.g. Lycett et al., 2004; Thiry, 2004) perspectives on programme management that are respectively performance and strategy oriented. As described in Section 5.2.1, the programme management literature suggests that the performance focus of project management and the more strategic focus of programme management are complementary (Pellegrinelli, 2011; Thiry, 2002). In addition, the case of Room for the River illustrates that it is worthwhile to combine these two foci both at the level of project and programme management, because they facilitate a collaborative approach between project and programme management. For example, the interview data suggest that this has led to more efficient realisation of projects as a result of facilitation of individual projects by the programme office and coordinated learning between projects (section 5.4.6).

However, despite the collaborative approach it should be noted that tensions occurred, at times, between project and programme management. For example, the performance focus of the programme management implied that the strict boundary conditions were not flexible after these were agreed upon by the key stakeholders. This tension manifested itself, for example, when regional stakeholders along the river IJssel were arguing (during the design phase) for accommodating a peak discharge of 18,000m3/s instead of the programme's objective of 16,000m3/s as this would further postpone future measures in their region. This led to the commissioning of an additional analysis of the effectiveness of the proposed measures by the national government and caused a delay of several months for the flood discharge channel project Veessen-Wapenveld. Furthermore, several project managers indicated that whilst the 'justification cycle' created a basis for facilitation by the central programme office, it also caused an administrative burden to appropriately report the progress of the projects with scarce resources available. Similar issues of inflexibility and administrative burdens are described in the programme management literature (e.g. Lycett et al., 2004; Platje and Seidel, 1993). However, an important difference with the tensions described in the programme management literature is that there was no competition for resources between Room for the River projects as the programme budget as the programme's flood protection objective prevailed over all other objectives and the project objectives and budgets were established in a initiation process in which all key stakeholders engaged intensively.

Overall, it can be concluded that the tensions between project and programme management were limited to such an extent that they have, so far, not adversely affected the performance of the programme. It should be noted that the programme success cannot be attributed to effective programme management alone, but also to contextual changes that positively affected the programme success. For example, the Room for the River programme was initiated and budgeted before the economic crisis whilst the construction contracts were signed during the economic crisis. According to several insiders, this resulted in lower than anticipated costs for the realisation of the river widening projects. Also, it was suggested that the competencies of the staff involved with the management of the projects and programme were amongst the best available in the organisations involved.

5.5.2 Reflection

The research findings demonstrate that particularly programme adaptation contributed significantly to the performance programme management effectiveness of the Room for the River programme. This finding is conform the suggestions of others that have pointed out that that programme success depends significantly on the ability to cope with contextual changes (Ritson et al., 2011; Shao et al., 2012). As illustrated by the Room for the River case study, this ability to adapt is determined by the design and, subsequently, the management of a programme. Programmes need to acknowledge the complexity of their context (Ritson et al., 2011). Dealing with complexity is, despite exceptions (e.g. Aritua et al., 2009; Hertogh and Westerveld, 2010; Ritson et al., 2011), new to research about programme management, but more commonly embedded in the literature about governance (e.g. Folke et al., 2005; Olsson et al., 2006; Teisman, 2005) and organisations (e.g. Boisot and McKelvey, 2010; Gomez and Jones, 2000). For example, adaptive governance research has described 'management as learning' approaches that consist of exploring problems and uncertainties, deliberating alternative solutions and reframing problems and solutions (van Herk et al., 2011b) and iterative re-evaluation of the fit-for-purpose of applied management approaches under different contextual conditions (Rijke et al., 2012a). Alternatively, organisational research has, for example, developed insights in the nature and properties of conventions within contexts and how these evolve over time to change the 'deep institutional' structures that are at the heart of organisations (Gomez and Jones, 2000). I therefore recommend further investigation of the applicability of the insights about dealing with complexity from, for example, the governance and organisational literature for programme management.

In addition, the findings illustrate the importance of stakeholder collaboration throughout the Room for the River programme, which confirms the findings of others that point out the benefits of stakeholder engagement in terms of, for example, legitimacy, knowledge management, early signalling of potential problems (e.g. Beringer et al., 2013; Hertogh et al., 2008; Hertogh and Westerveld, 2010). Establishing a stable programme management approach that is able to respond to uncertain and changing conditions can be considered one of the most challenging aspects of programme management (Davies and Mackenzie, 2013; Sanderson, 2012). Without clearly defined boundary conditions of the programme design, particularly the clearly defined roles and responsibilities, it would be uncertain whether stakeholder engagement had a positive effect on the programme management performance (Beringer et al., 2013). In addition, other research suggests that the level of autonomy of the programme management office within its hosting organisation Rijkswaterstaat has a positive effect on its ability to manage the programme, as interventions from the hosting organisation undermine the position of programme managers and, thus, negatively affect stakeholder collaboration (Jonas, 2010).

We note that the six attributes for effective programme management that were identified from the literature are contributing differently to the programme management performance, but that they cannot be seen in isolation of each other. For example, the three attributes of the programme design (i.e. programme vision, priority focus, planning framework) that were established during the initiation stage were highly influential on the management processes that took place during the consecutive stages of the Room for the River programme (see also Heising, 2012; Reiss et al., 2006; Shehu and Akintoye, 2009). The research data suggest that particularly stakeholder collaboration and programme adaptation are highly interrelated factors that reinforce each other (section 5.4.6). As such, it can be concluded that stakeholder collaboration and programme adaptation have a recursive relationship and both factors cannot be considered in isolation of each other (see also Feldman and Orlikowski, 2011). Especially because the interplay between both factors were considered at the heart of successful balancing between a strategic and performance oriented focus of the management of the individual projects and the programme as a whole, I suggest to further investigate how stakeholder collaboration and programme adaptation can reinforce each other effectively within megaprojects such as Room for the River.

As this research is based on a single case study, it is impossible to derive generally applicable conclusions from the analysis. In the light of complexity, it is uncertain how the lessons of the analysis of the Room for the River programme will translate to other megaprojects. Moreover, it is impossible to reconstruct the exact conditions in which the Room for the River programme emerged and evolved. However, as complexity implies a combination of both chaos and order (Boisot and McKelvey, 2010), it is likely that the findings could provide, at least some, relevant insights for other megaprojects. I therefore recommend exploiting the lessons from the Room for the River case study as a starting point for developing appropriate programme management approaches of new megaprojects. Applicability of the lessons from the analysis would depend on the nature of the megaprojects and their context. For example, it would be worthwhile to investigate if the lessons are applicable to programmes with other programme management styles than the combination of a goal-oriented and service centre programme management styles that was applied in the Room for the River programme. Also, it would be interesting to investigate the implications of a lesser degree of autonomy of individual projects, such as arguably in megaprojects of road, rail and powerplant infrastructure, on the capacity and organisation of programme adaptation.

5.6 Conclusion

This paper presents the findings of a study of how the programme management processes in the Room for the River programme have evolved. Using six attributes for effective programme management that were identified from the literature, it was identified that, during the initiation stage, the Room for the River established: 1) a *clear programme vision* for flood risk protection through river widening that was widely supported by all relevant stakeholders; 2) a *clear priority focus* that provided opportunities to connect stakeholder ambitions to the overall programme objectives though combining the flood protection objective with a secondary objective for achieving spatial quality; and 3) a *transparent programme planning framework* that outlined the boundary conditions and roles of the stakeholders and that was based on a steering philosophy of "controlled trust". Furthermore, three key management processes were identified for the consecutive stages of the programme: 4) *programme governance* involving internal and external stakeholders that matches the vision, priority focus and planning framework of the programme to enhance the legitimacy and quality of the programme and its projects; 5) *appropriate programme coordination* to monitor progress of intermediate milestones and management performance and, if needed, assist projects in achieving their objectives; and 6) *programme adaptation* to adjust the programme's organisation or plans and designs to the context of the individual projects and the programme as a whole.

The analysis of the case study shows that the focus of the programme management office and the individual project teams differed from what is traditionally expected from the programme management literature. As a result of the programme design, which was set up to support decentralised decision making within centralised boundary conditions, both the programme and project management have been balancing their focus between creating strategic value (through stakeholder collaboration within the local and national context) and achieving high performance (through coordination). This has particularly benefited the collaboration between project and programme management and, as a consequence, the capacity of the programme and projects to adapt to new insights and changing contextual conditions. The relevance of the analysis may be relevant for other megaprojects within and beyond the domain of flood risk management as unfortunately cost and time overruns are unfortunately common for such projects. I suggest that the results of the analysis are used as a starting point for developing appropriate programme management approaches of new megaprojects. However, applicability of the lessons from the analysis would obviously depend on the nature of these megaprojects and their context.

CHAPTER SIX

Governance for strategic planning and delivery of adaptation

Planning for and delivery of adaptation.

In addition to the previous chapter, this chapter addresses the 'how' question by connecting governance for the strategic planning for adaptation with governance for the delivery of adaptation. Based on a comparative study of adaptation in the water sectors in the Netherlands and Australia, this chapter describes that the uptake of planned adaptation action can be stimulated through reinforcing connections between the governance for strategic planning and the governance of delivery of adaptation projects.

6. Governance for strategic planning and delivery of adaptation

This chapter is adapted from:

Rijke, J., Brown, R., van Herk, S., Zevenbergen, C. (under review) Governance for strategic planning and delivery of adaptation. *Journal of Environmental Planning and Management*.

Abstract

This chapter discusses how governance for strategic planning and delivery of adaptation can be aligned more effectively in order to successfully realise adaptation action. Previous research on governance of adaptation has focused predominantly on strategic planning for adaptation and has largely overlooked the delivery of adaptation in practice. Meanwhile, there is a gap between aspirations for adaptive water management systems and the realisation thereof. Using two cases that are globally regarded as leading in terms of implementing innovative water management approaches, I have analysed the coming about of adaptation action by investigating the interactions between the governance for strategic planning and the governance for the delivery of adaptation. These cases show that governance of strategic planning can enhance delivery through creating the conditions that are needed to deliver adaptation action effectively, including stakeholder support, a broad knowledge base and an allocated investment budget for the realisation of adaptation action. Conversely, both cases show that governance of delivery can be influential for strategic planning of new adaptation actions through knowledge and relationships that are developed for the realisation of adaptation action. Hence, I conclude that governance for strategic planning and also for the delivery of adaptation action can be mutually reinforcing. As a consequence, scholarship related to the governance of adaptation would benefit from refocusing its current emphasis on strategic planning towards an approach that also incorporates a lens for implementation in order to turn aspirations into reality.

6.1 Introduction

Adaptation to climate change is commonly referred to as a governance challenge (e.g. Adger et al., 2009; Folke, 2006; OECD, 2011). With regard to water management, the focus of this paper, this governance challenge can be broken down into several parts. Firstly, the practical implementation of available innovative technologies and of the knowledge required to develop adaptive water management systems is slow (Harding, 2006; Mitchell, 2006). Secondly, it is nowadays frequently suggested that new modes of governance are needed that are effective under conditions of high complexity and uncertainty (e.g. Folke et al., 2005; OECD, 2011; Zevenbergen et al., 2012). These approaches would involve, for example, multiple disciplines, multiple government levels, the community, the private sector and academia. Adger and colleagues summarise these governance challenges by suggesting that *"adaptation to climate change is limited by the values, perceptions, processes and power structures within society"* (Adger et al., 2009, p.349).

These governance challenges are broadly being addressed by research related to various theoretical backgrounds. For example, from the social-ecological systems perspective, the concept of adaptive governance has been developed to deal with the complexity and uncertainty that are inherent in social-ecological systems (systems in which ecosystems and social systems co-evolve). Adaptive governance is about enhancing the capacity of individuals, groups and institutions to anticipate long-term change (e.g. climate change, population growth), respond to immediate shocks (e.g. drought, flooding) and recover from such shocks (e.g. Brunner et al., 2005; Dietz et al., 2003; Folke et al., 2005). This is being considered as a key ingredient for establishing resilient systems, where resilience can be defined as the capacity of a system: (i) to absorb shocks while maintaining function (Holling, 1973); (ii) for adaptation and learning (Folke, 2006; Gunderson, 1999; Olsson et al., 2006); and (iii) for renewal and reorganisation following disturbance (Gunderson and Holling, 2002).

In addition, from the socio-technical systems perspective, the concept of transition management has been developed to establish systemic change in socio-technical systems (systems in which society and technology co-evolve). In relation to water management, transition management scholarship aims to support an increase in the sustainability or resilience of existing water systems (Brown et al., 2009b; Pahl-Wostl et al., 2010; van der Brugge and Rotmans, 2007). This paper takes a combined perspective on governance of adaptation by defining its purpose as: (1) to successfully anticipate systemic change; and (2) to deliberately stimulate systemic change.

Recent studies in, for example, Australia, the UK and Scandinavia, have shown that adaptation research findings are often not being adopted in practice (Brown et al., 2011; Klein and Juhola, 2013). It is suggested that slow uptake of adaptation research into practice can be attributed to a dominant research focus on system performance which is currently neglecting the perspective of decision-makers and the role of agency (Klein and Juhola, 2013). Moreover, recent research has pointed out that the considerable research efforts on adaptation mainly focuses on intentions to adapt rather than on real adaptation actions in practice (Berrang-Ford et al., 2011). This suggests that governance for the delivery of adaptation action is largely overlooked by adaptation research.

This paper contributes to the existing body of adaptation research by connecting governance for strategic planning for adaptation and governance for delivery of adaptation action into practice. I draw from empirical insights from two cases of adaptation in water management in the Netherlands and Australia. These cases have been selected because their relative success in terms of uptake of adaptation action suggests that strategic planning pushes delivery effectively. Based on the assumption that this is indeed the case, I test the hypothesis that governance for strategic planning and governance of delivery of adaptation can reinforce each other and should therefore be aligned in order to stimulate adaptation action.

6.2 Theoretical background: Governance for strategic planning and delivery of adaptation

The IPCC defines adaptation as *"the adjustment in natural or human systems in response to actual or expected climatic stimuli or their effects, which moderates harm or exploits beneficial opportunities"* (Parry et al., 2007, p.6). Adaptation is a transformational process that is induced by self-organisation and/or deliberate planning (Folke, 2006; Moser and Ekstrom, 2010). With regard to deliberately planned adaptation, the focus of this paper, adaptation can be considered as a continuous cycle of activities for understanding the need for adaptation, planning for adaptation and managing adaptation action (Moser and Ekstrom, 2010; Figure 6.1). Because implementation of adaptation in practice is the primary concern of this paper, I distinguish between strategic planning for adaptation and delivery of adaptation. In this paper, strategic planning for adaptation refers to activities that relate to understanding the need for adaptation and planning for adaptation action. Delivery of adaptation refers in this paper to managing adaptation action in practice after a particular action has been selected. Based on a review of the literature about adaptive governance, transition management and project/programme management, this Section discusses the main elements and differences between strategic planning and delivery of adaptation: aim, focus, measure of effectiveness, purpose of learning, purpose of collaboration (Table 6.1).

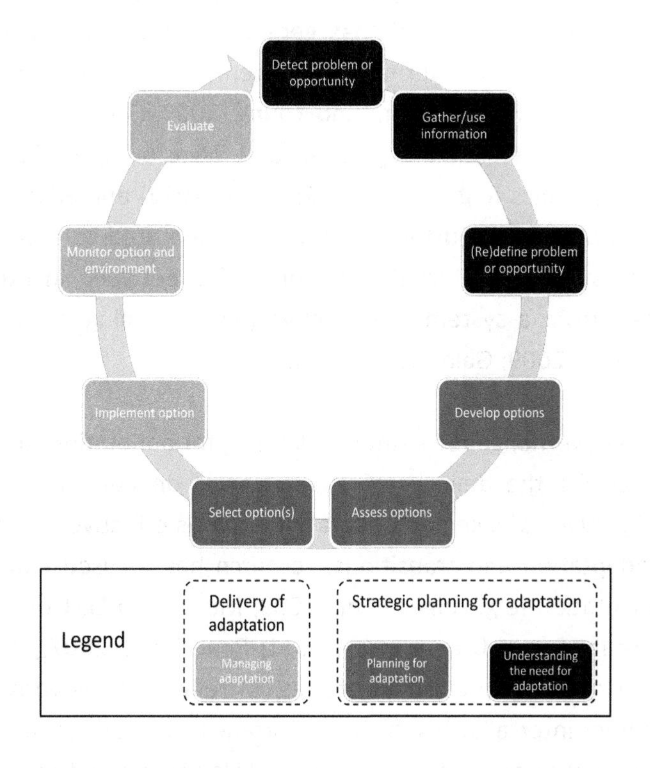

Figure 6.1 **The adaptation process: strategic planning and delivery of adaptation (adapted from Moser and Ekstrom, 2010)**

Governance for strategic planning for adaptation involves the organisation of processes from detection of problems and/or opportunities to the selection of options that address these (Figure 1). A review of the strategic planning literature recently revealed that strategic planning is generally considered a systematic, step-wise approach to strategy formulation, implementation, and control (Wolf and Floyd, 2013). However, the actual pattern of actions and decisions is not a product of strategic plans, because it is in reality influenced by emergent factors that are not anticipated in the plans (Mintzberg, 1994; Mintzberg and Waters, 1985). As such, the main aim of strategic planning for adaptation is to develop understanding about the need for adaptation and to develop adaptation options. Adaptation research is using a systems approach to assess the fit between the institutions and the biophysical and social domains in which they operate for understanding of the effectiveness and robustness of governance under changing conditions (Galaz et al., 2008; Rijke et al., 2012a; Young and Underdal, 1997). Research related to assessing the aforementioned fit has generated insight about how to improve the functionality of governance arrangements for different contexts in terms of ecosystems (e.g. Ekstrom and Young, 2009; Shkaruba and Kireyeu, 2012), technological systems (e.g. Elzen and Wieczorek, 2005; Lieberherr, 2011), social dynamics (e.g. Ebbin, 2002; Meek, 2012) and adaptation pathways (Rijke et al., 2013). Furthermore, it has attempted to enhance the fit through understanding different types of misfits between institutions and their contexts from a systems perspective (e.g. Cumming et al., 2006; Ekstrom and Young, 2009; Galaz et al., 2008).

In contrast to governance for strategic planning for adaptation, the main aim of governance for the delivery of adaptation is on performing tasks and achieving objectives of selected adaptation options effectively. In relation to governing adaptation and transitions, research has focused extensively on limits for adaptation (e.g. Adger et al., 2009; Moser and Ekstrom, 2010) and barriers to change (e.g. Brown et al., 2011; Pahl-Wostl, 2007), and thus on challenges to meet adaptation objectives. In relation to governance these barriers include, inter alia: insufficient awareness and understanding about the need for adaptation; limited resources; skills and competencies to adapt; fragmented institutional arrangements; limiting regulatory environments for technological innovation; and, ineffective leadership. Accordingly, the research about governance of adaptation has, so far, mainly focused on experimentation to overcome these barriers (e.g. Farrelly and Brown, 2011; van Herk et al., 2011a). Because implementation of adaptation is in many cases insufficiently put in practice (Brown et al., 2011; Klein and Juhola, 2013), it appears that the literature on governance of adaptation has not evolved to support adaptation action effectively.

Considering that *"policy cannot always mandate what matters to outcomes at the local level"* and that *"policy-directed change ultimately is a problem of the smallest unit"* (McLaughlin, 1987; p.171), it seems logical to investigate the implementation of adaptation action at the level of the projects through which implementation occurs. As this research focuses on the organisational aspects of adaptation action, I turn to the domain of project management for theoretical guidance for actual implementation of adaptation (e.g. Cleland and Ireland, 2002; Kerzner, 2009; Turner, 1999). As cost overruns and time delays are commonplace for large scale infrastructure projects (Flyvbjerg, 2007; Flyvbjerg et al., 2003), this research domain focuses primarily on the effective achievement of project objectives, which is often measured through project effectiveness, efficiency and stakeholder satisfaction (Shao et al., 2012; Shehu and Akintoye, 2009) and, sometimes, value creation for the organisations involved (Murray-Webster and Thiry, 2000; Thiry, 2002, 2004). The project management literature has, for example, aimed to enhance project success by providing attributes for adequate project design, desired competencies of project managers, and effective interaction between the project management and the context in which projects take place (Hertogh and Westerveld, 2010; Turner, 1999). It should be noted, however, that the scholarship of project and programme management is not connected to the governance of adaptation research.

Stakeholder collaboration and learning play important roles in both strategic planning and delivery of adaptation. In relation to strategic planning for adaptation, the concepts of 'management as learning' and multi-level governance take a central position in order to make decisions that are based on a broad knowledge set to avoid unexpected or undesirable outcomes (Edelenbos and Klijn, 2006; Pahl-Wostl, 2007). At the strategic level, collaborative governance approaches are considered necessary particularly for enriching strategies, policies and plans by integrating technical, legislative and procedural knowledge from different disciplinary and organisational perspectives (Edelenbos and Klijn, 2006; van Herk et al., 2011a). As such, stakeholder collaboration is a prerequisite for joint learning, which in itself is considered a key ingredient for adaptation of complex systems (Folke et al., 2005; Ison et al., 2007). In relation to strategic planning, collaborative learning plays a key role in transforming systems and reframing guiding assumptions about policy goals, problems, system boundaries and potential solutions (Armitage et al., 2008; Pahl-Wostl, 2009). The project management literature is increasingly adopting collaborative and adaptive management approaches in which learning plays a key role (Hertogh and Westerveld, 2010; Rijke et al., in press; Ritson et al., 2011). In this light, collaborative approaches that balance between project control (e.g. control of budget, time, risk) and stakeholder engagement are considered important for adjusting routines and project management approaches to changing circumstances and consequently achieve desired project outputs (Hertogh and Westerveld, 2010; Rijke et al., in press; van Herk et al., in press).

Table 6.1 Governance for strategic planning and delivery of adaptation

	Governance for strategic planning of adaptation	**Governance for delivery of adaptation**
Main aim	Understanding the need and developing options for adaptation	Implementing adaptation options
Focus	System focused: Maintaining or improving system functionality across spatial, temporal and institutional scales	Actor-oriented: Achieving objectives and tasks
Measure of effectiveness	Spatial, temporal and functional 'fit' between governance and biophysical and social systems	Project effectiveness, efficiency and stakeholder satisfaction
Purpose of learning	To adapt to changing risks or new opportunities	To achieve objectives and perform tasks effectively
Purpose of collaboration	Control vs. interaction: Coordination of resources versus knowledge exchange	Control vs. interaction: Project control versus stakeholder engagement

Table 6.1 summarises the key differences for governance for strategic planning and governance of delivery of adaptation. The key activities during the planning and delivery stages of adaptation are carried out by different sets of actors and, thus, by different governance configurations: policy makers and scientists are mainly involved with the strategic planning for adaptation;

whilst practitioners (in the public and private sector) and decision makers perform key roles during the delivery of adaptation. As it appears difficult to transform strategic planning into delivery of adaptation action, this paper aims to demonstrate how governance for strategic planning and delivery of adaptation can be aligned more effectively.

6.3 Research approach

This paper demonstrates that governance for strategic planning and delivery of adaptation should be considered in combination in order to achieve adaptation in practice. This paper draws on two cases of adaptation in the context of water management: river widening as a means to adapt to increased flood risks in the Netherlands; and stormwater harvesting and reuse as a means to adapt to drought in Australia. Both cases are selected because they are considered relatively successful in terms of the delivery of the projects (Farrelly and Brown, 2011; Rijke et al., 2012c) and because they are both globally recognised as being advanced in strategic planning for the innovative water management approaches adopted (Howe and Mitchell, 2011; Warner et al., 2013). Therefore, I assume that strategic planning is pushing delivery of adaptation relatively effectively in each case. Furthermore, both cases are selected as they are assumed to represent two different adaptation pathways: the case of stormwater harvesting and reuse in Australia represents an incremental adaptation pathway in which informal coalitions of actors collaborate to drive and upscale experimental applications (Farrelly and Brown, 2011; Rijke et al., 2013), whereas the case of river widening in the Netherlands represents systemic transformation in which the implementation of river widening is ultimately formalised in a multi-billion euro investment programme (Rijke et al., 2012c). This difference provides the opportunity to test the hypothesis that governance for strategic planning and governance for delivery of adaptation can be mutually reinforcing, even for different adaptation pathways.

For both cases, I analyse how adaptation action has come about by describing separately the governance for the strategic planning and the delivery of adaptation. Subsequently, I investigate for each case how the governance for strategic planning and the delivery of adaptation action have interacted. In particular, I examine when the focus of strategic planning shifts towards the delivery of adaptation action. Moreover, I examine the role of relationships between actors, knowledge and financial resources during both stages, because these factors give insight to learning, collaboration and effectiveness of both stages of adaptation (see Table 6.1). Based on the comparative insights of the two cases I analyse if governance for strategic planning and governance for delivery of adaptation are actually mutually reinforcing.

Table 6.2 Summary of research methods for the two cases

Method	Case: River widening in the Netherlands	Case: Stormwater harvesting in Australia
Document analysis	Policy documents, progress reports, evaluations, media coverage	Policy documents, media coverage
Literature study	Relevant scientific publications	Relevant scientific publications
Semi-structured face-to-face interviews	n = 55 respondents	n = 90 respondents
Validation interviews	-	n = 20 respondents
Validation workshops	2x, total n = 33 participants	4x, total n = 93 participants
Observation at training sessions and political conferences	6x, total n = 415 participants	-

Multiple sources of evidence were used to analyse the cases, including data collected through face-to-face interviews, document analysis and literature study (Table 6.2). For both cases, the sets of interviewees were selected to represent the context of governance of the respective adaptation action. They included a broad range of stakeholders that together represented the key organisations involved and consisted of key decision makers and individuals in senior advisory roles that were involved with the early development of the innovative concepts and their colleagues who are presently involved with policy making, regulation and the planning, realisation, operation and maintenance of projects. Furthermore, they had different disciplinary backgrounds and professional roles, such as engineering, spatial planning, ecology, economics and politics. Individuals who contested the use of river widening and stormwater harvesting and reuse were also interviewed to capture a balanced view including potentially conflicting governance strategies.

All interviews were semi-structured around the theme of adaptation action (i.e. river widening/stormwater harvesting and reuse) to acquire the interviewees' personal insights about the delivery of adaptation action and to provide the opportunity for new ideas to emerge during the interviews. The recurrent interview questions during each of the interviews related to: 1) the associated values of adaptation through respectively river widening and stormwater harvesting and reuse and alternative adaptation options; 2) the drivers and barriers for strategic planning for adaptation action; 3) the drivers and barriers for the delivery of adaptation action; 4) the strengths and weaknesses of these personal and organisational strategies to plan and deliver adaptation action. The personal insights of the interviewees were transformed into case study narratives by structuring the data around the recurrent questions, identification of dominant perceptions and extensive validation thereof.

Following the first stage of data analysis, an extensive validation process was undertaken to test the research findings (Table 2). Validation included a review of policy reports, legislation, regulation and media documentation to support, specify and/or contradict interviewee interpretations. Moreover, validation interviews (only for the stormwater case) and validation workshops (both cases) were conducted with representatives of key stakeholder groups who were previously interviewed and with individuals with an overview about water governance in both the Australian and Dutch contexts. In addition, validation of the Dutch case study occurred through observations at three stakeholder management training sessions for three project teams for river widening measures in the Dutch national Room for the River programme, two political conferences for national, regional and local decision makers from the Room for the River project areas, and one conference for professionals within Rijkswaterstaat, which hosts the programme management office of the Room for the River programme.

6.4 Results

In this section, the case study findings are described. Firstly, the adaptation action that has been implemented is summarised. Secondly, the governance for strategic planning for adaptation action is described. Thirdly, the governance of the delivery of adaptation action is described. And finally, how the strategic planning and adaptation action are interacting is considered.

6.4.1 River widening in the Netherlands

The case of river widening in the Netherlands illustrates that strategic planning for adaptation can effectively lead to system wide implementation of adaptation action. It also shows that the knowledge and relationships that are developed for the execution of adaptation action can be beneficial for strategic planning of adaptation actions in the near future.

6.4.1.1 Adaptation action

Since approximately 1100 AD, the areas in the Netherlands along the branches of the rivers Rhine and Meuse have gradually been developed for agricultural production and urban development (Huisman, 2004). This has reduced space for the flows of the rivers and required repeated heightening of dikes to protect against flooding. However, two near-miss floods in 1993 and 1995 catalysed the emergence of a new approach to adaptation as they triggered a two-stage response of immediate dike improvements where the flood safety standard was not met and the development of a new policy for river widening to increase the flood safety standard. The latter was established in 1996 as the Room for the River Policy Directive that was the first policy framework to replace the traditional approach to adaptation by repeated strengthening of flood defences. Within the scope of the Room for the River Policy Directive, the Dutch Government laid out a programmed approach for reducing flood risk along the river Meuse (Maaswerken programme) and the river Rhine and its main river branches (Room for the River programme).

The €2.4 billion Room for the River programme constituted the largest part of the proposed adaptation action within this policy framework. This programme initially consisted of 39 projects (later reduced to 34 projects) that create more room to accommodate river discharge using measures such as floodplain excavation, peak discharge channels and dike relocation. With this, the Room for the River programme will achieve an increased flood protection standard of a 1:1250 year design discharge capacity. By 31 December 2012, the plans for 97% of the programme budget were completed and construction contracts were signed for 70% of the programme budget. The programme is on schedule to achieve its objectives within budget and with only minor delays (PDR, 2013) in a manner that is, overall, satisfactory to all the stakeholders involved (Rijke et al., 2012c; van Twist et al., 2011b).

6.4.1.2 Governance for strategic planning of river widening

Since the 1970s, a combination of factors contributed to the emergence of an alternative approach to adaptation that consisted of river restoration and creating more space for rivers for accommodating peak discharges: a scope widening of water management; community resistance against dike strengthening; experimentation to develop alternatives for dike strengthening; and a sudden political momentum after the two near-miss floods in 1993 and 1995 (Table 6.3; see also van der Brugge et al., 2005; van Heezik, 2007).

Table 6.3 Key drivers that contributed to the emergence of river widening for flood protection (1970s-1995)

Driver	Explanation
Scope widening of water management	Since the 1970s, increased environmentalism started to challenge the 'control paradigm'. The first time this happened at a national scale, was during the public debate that led to a significant alteration of the original construction plan of the Eastern Scheldt storm surge barrier in 1974. Irreversible ecological damage of the salt water environment that would be caused by the closing of the Eastern Scheldt estuary was avoided through the construction of moveable gates that would only be closed under extreme circumstances (Knoester et al., 1984). Later in the 1980s, the technocratic control paradigm was further challenged by the emergence of the concept of "integrated water management" that originally aimed to avoid conflicts between different uses of water resources through improved coordination (Saeijs, 1991).
Resistance against dike strengthening	In the 1970s and 1980s, many dike strengthening projects were delayed due to public resistance and legal procedures. Opponents particularly argued that flood safety was the only factor that was taken into consideration for the planning of these projects and that landscape, environmental and cultural values of riversides and surrounding areas were overlooked.
Experimentation with alternatives for dike strengthening	Driven by the resistance against dike strengthening, several small scale experiments were undertaken from the 1980s onwards to develop alternatives that took landscape, environmental and cultural values into account. In addition, the studies 'Plan Ooievaar' (de Bruin et al., 1987) and 'Living Rivers' (WNF, 1992) showed that integrated approaches that included river widening could reduce flood risks whilst improving environmental values.
Political momentum after two near-miss floods	The near-miss floods in 1993 and 1995 created a broadly shared sense of urgency to take adaptation action. Particularly the near-miss flood in 1995 which created great political momentum to take action, as it led to the evacuation of 250,000 people and 1 million cattle.

Within this context, the Room for the River programme was initiated in 2000 and formally established in 2006. In the context of the political momentum that emerged after the two near-miss floods, it could be concluded that river widening was a deliberately planned adaptation action in the Netherlands. The governance for strategic planning of river widening measures operated within the boundaries of the Room for the River Policy Directive of 1996. The governance processes that took place during the initiation phase of the Room for the River programme were principally aimed at establishing a legitimate set of flood protection measures. In the design of the Room for the River programme, the overall performance of the flood risk management system and the programme as a whole prevailed over individual projects. The river widening measures were selected through a systems approach for flood protection that considered the complete system of the river Rhine in the Netherlands. The selection process involved active engagement of local politicians to select legitimate and feasible measures and consulted other stakeholders to consider knowledge about the impact and opportunities on their respective interests (e.g. agriculture, business, nature, recreation) in the selection process.

The completion of the initiation phase of the Room for the River programme was demarcated by the establishment of the Policy Decision Room for the River (PKB Ruimte voor de Rivier; see also ten Heuvelhof et al., 2007). This policy document captured the new vision for flood risk management that was based on river widening rather than strengthening dikes and explicitly included spatial planning. Accordingly, it set out the programme's formal objectives of: 1) improving safety against flooding of riverine areas of the Rivers Rhine, Meuse, Waal, IJssel and Lek by accommodating a discharge capacity of 16.000m3/s; 2) contributing to the improvement of the spatial quality of the riverine area. The secondary objective of enhancing the spatial quality of the river area was included in the PKB as a strategy to involve regional stakeholders and thus increase the legitimacy. It provided the opportunity for regional stakeholders to connect local ambitions to the flood safety measures. Furthermore, the PKB provided a planning framework for the implementation of the new vision for flood risk management by setting out the boundary conditions and performance criteria for each of the selected river widening projects, and the roles and responsibilities of the key stakeholders that would be involved in the planning, design and realisation of these projects. The PKB was jointly signed by the national government, provinces, waterboards, and municipalities involved to secure commitment to execute the Room for the River programme according to plan.

As such, governance for strategic planning of the Room for the River programme provided a legitimate set of boundary conditions for the governance during the planning, design and realisation phases of the adaptation measures.

6.4.1.3 Governance of delivery of river widening

A new multi-level governance approach was used for the planning, design and realisation of the programme in which the regional and local governments had a lead role in the design and decision making processes, whilst a programme office assigned by the national government monitored the progress, performance and quality of the projects and facilitated where necessary (see Rijke et al., in press). This strategy enabled local stakeholders to identify opportunities to connect local ambitions and signal any emerging issues in the local context. In addition, the centralised programme office aimed to coordinate the programme in a way that it delivered the initial programme objectives whilst adapting to changing conditions in project and programme contexts. For this, it coordinated learning in the programme through: 1) a pre-set procedure for milestone management as a basis for regular evaluation of project performance in terms of hydraulic performance, spatial quality, legal procedures, soil, integrated design, budget and risk management; and 2) a consistent balance between project control, process (or stakeholder) management and technical management throughout the management of each individual project and the programme as a whole, as a basis for knowledge sharing between project teams and programme management. This combined management arrangement was instrumental in proactively signalling (potential) problems and taking remedial action. Furthermore, it enabled the programme management to transfer lessons learned across all projects within the programme and addressed overall political, policy and economic contexts. In addition, it enabled the coordination of financial resources at a programme level by cancelling several projects after other projects realised greater water level reduction than initially anticipated.

6.4.1.4 Interaction between strategic planning and adaptation action

The case of river widening in the Netherlands illustrates the differences between governance for strategic planning of adaptation and governance of the delivery of adaptation. As described above and summarised in Table 6.4, governance for strategic planning for river widening focused on establishing legitimacy and stakeholder commitment and developing a set of boundary conditions for the subsequent planning, design and construction of the adaptation measures. As such, it explored what would be realistic, feasible, legitimate and effective adaptation measures. The governance for the delivery of the adaptation measures built on the collaborative approach and relationships that were established during the initiation phase, but also explicitly focused on the effective and efficient delivery of the proposed adaptation measures.

Table 6.4 Summary of governance for strategic planning and governance of delivery of river widening

	Governance for strategic planning for river widening	Governance of delivery of river widening
Timing	Starting from 1970s, but sudden expansion of momentum after the near-miss floods in 1993 and 1995 which led to the initiation phase of Room for the River (2000 - 2006).	During the planning and design phase of Room for the River (2007 - 2012).
Relationships	Stakeholder engagement for creating support and commitment to collaborate with the realisation of river widening measures. Interactive selection of a set of measures to improve legitimacy and feasibility of the programme.	Task setting of organisations and individuals in individual projects and the programme as a whole. Community building amongst individuals with similar roles. High-level (political) interaction when projects faced issues that could not be solved otherwise.
Knowledge	System analysis to explore possible, preferential and feasible measures and opportunities to link local ambitions.	Monitoring of progress, quality and risks of individual projects and the programme as a whole. Facilitation of individual projects by the programme office through expert knowledge, guidelines and political influence.
Financial resources	Allocation of a programme budget.	Monitoring and half-yearly reporting of project and programme progress, budgets and risks. Re-allocation of budgets when unexpected high performance of measures made others abundant.

In the case of river widening in the Netherlands, strategic planning has successfully pushed for the implementation of adaptation actions by building a new policy vision and translating it into a programme for system wide adoption of river widening. Particularly during the initiation phase of the Room for the River programme (2000-2006), the policy vision was converted into a programme vision (flood risk reduction through river widening), priority focus (improving flood safety whilst contributing to spatial quality) and programme strategy (decentralised planning and decision making within centrally set and safeguarded boundary conditions). During the implementation of Room for the River, strategic partnerships were created with politicians, policy makers and regulators to secure programme progress. Through these partnerships, the programme has been informing contemporary water related policy making and governance approaches of future large infrastructure projects, such as the new national Flood Defense Programme (2011-2017) and the Delta Programme (2010-2014).

In addition, the delivery of the Room for the River programme has impacted future strategic planning activities through the development of policies and regulation, practical guidelines for the implementation of policies and the use of new tools (see also van Herk et al., 2013). Policy changes and precedents for the interpretation of new policies were created during the delivery of the Room for the River programme in order to achieve the programme's objectives. Examples of impact on policy include: policy for land use in outer marches; precedents for dyke requirements; regulation on soil and water quality; regulation for redevelopment of lakes; and nature-oriented planning. Furthermore, coordination of learning between river widening projects in the programme the implementation of policy in the future has resulted in the development of guidelines for soil movement planning, planning for spatial quality, groynes information systems, consistent information requirements for hydraulic, vegetation, landscape mapping and planning, and asset monitoring and maintenance protocols. Moreover, the decision making support tool ('Blokkendoos') that was used during the interactive selection process of the river widening measures in the programme is now being adapted for strategic planning of future flood risk management in the Delta Programme.

Hence, it can be concluded that governance for strategic planning and governance for delivery of adaptation have been mutually reinforcing for river widening in the Netherlands.

6.4.2 Stormwater harvesting and reuse in Australia

The case of stormwater harvesting and reuse in Australia illustrates that the distinction between governance for strategic planning and governance of delivery of adaptation is not always clear. It also shows that a deliberate learning agenda of a professional network of advocates can bring together both elements of governance of adaptation.

6.4.2.1 Adaptation action

Traditionally the water supply of Australian cities relies on highly centralised 'big pipes-in big pipes-out' infrastructure in which water is supplied from catchments and water reservoirs that are often located outside the metropolitan area. A prolonged drought (2000-2009) has significantly affected Australian water resources and forced State governments to restrict water use, which in turn has made water management a hot political topic with much media attention. In order to secure water supplies, each of Australia's State capital cities has augmented its water resources through rainfall-independent desalination plants and wastewater recycling schemes that were extensions of the existing centralised water supply infrastructure. In addition, investments are made in rain-dependent measures such as rainwater tanks and stormwater harvesting and reuse schemes.

Implementation of stormwater harvesting and reuse schemes is largely funded through the Australian government's National Urban Water and Desalination Plan, which allocated AU$200 million for urban stormwater harvesting and reuse projects during 2009 and 2010 (50% co-funding by e.g. state and/or local governments required). This has triggered large scale investments in stormwater harvesting and reuse projects (for an overview, see Australian Government, 2013), particularly in Adelaide where geological conditions enabled the large scale implementation of Aquifer Storage and Reuse schemes. Implementation of stormwater harvesting and reuse schemes is largely taking place in a fragmented manner, despite the fact that the majority of these schemes are being funded through one national fund. With urban drainage traditionally being a responsibility of local government, stormwater harvesting and reuse innovations are mostly driven by local governments, and in some cases water companies, who aim to reduce water demand for the irrigation of public space or other non-potable uses. The efforts to realise stormwater harvesting schemes are therefore scattered amongst a large number of organisations, because the metropolitan areas of the large Australian cities comprise a large number of jurisdictions (e.g. the metropolitan areas of Sydney, Melbourne and Adelaide consist of respectively 41, 38 and 19 municipalities).

6.4.2.2 Governance for strategic planning for stormwater harvesting and reuse

Stormwater harvesting and reuse takes a central part of a vision for water sensitive cities (WSCs). A WSC is considered to be adaptive and resilient to broadscale change (Wong and Brown, 2009). A WSC would achieve this through planning for diverse and flexible water sources (e.g. dams, desalination, water grids and stormwater harvesting), incorporating Water Sensitive Urban Design (WSUD) for drought and flood mitigation, environmental protection and low carbon urban water services in the planning system, and enabling social and institutional capacity for sustainable water management (see also Wong and Brown, 2009). A WSC is being considered as the outcome of amongst other things, WSUD processes that integrate water management and urban design and planning.

The emergence of stormwater harvesting and reuse is driven by a combination of factors, including an ongoing paradigm shift towards WSUD in which stormwater management and urban planning/design are increasingly integrated, experimentation with alternative water resources, increasing sense of urgency to take action triggered by the persistent drought and a national water reform agenda (Table 6.5).

Within this context, individuals involved with the strategic planning for stormwater harvesting and reuse made use of the ongoing paradigm shift towards WSUD. The peak of the drought in 2006 triggered state governments to redirect their resources and attention to mitigating the impact of the drought. As a result, the focus of WSUD has shifted from waterway health to water security. The improved inter-disciplinary connections between water management and urban planning and design that were established by the activities to improve stormwater quality management were also needed to use open space for harvesting and reuse of stormwater. As

Table 6.5 Key drivers that contributed to the emergence of stormwater harvesting and reuse in Australia (early 1990s-2006)

Driver	Explanation
Paradigm shift towards integration of stormwater management and urban planning/design.	Since the early 1990s, a network of individuals in state government, local government, academia and engineering consulting with a common vision for at-source stormwater treatment has emerged and strengthened. This network attempted to influence urban water policy and stormwater management practice to improve waterway health through collaborative research, experimentation (e.g. Lynbrook Estate) and capacity building. These activities enhanced a paradigm shift towards integration between water management and other disciplines such as urban planning and design and environmental management (see also Brown et al., 2013).
Experimentation with alternative water resources	Experimentation with stormwater harvesting and reuse increased the confidence about the quality, technical feasibility and financial viability since the early 1990s. Driven by a variety of factors including a need for flood mitigation, an environmental protection agenda and a vision for sustainable water supply gave rise to the development of the first Aquifer Storage and Reuse in Adelaide in which stormwater was harvested and used for irrigational purposes. Also in other states, a number of experiments were conducted across Australia with on-site treatment and reuse of stormwater, greywater and wastewater to explore how the impact of the drought could be mitigated (see Farrelly and Brown, 2011).
Persistent and increasing sense of urgency to take adaptation action to mitigate the impact of drought	Due to lengthy dry climatic conditions between 2000 and 2009, there was an increase in public concern to increasing water scarcity. Images of significantly dropped water levels in the drinking water reservoirs for e.g. Sydney and Melbourne maintained a persistent sense of urgency amongst the public, media and politics to take adaptation action. In Adelaide, where there was an immediate threat of running out of water supplies at the peak of the drought in 2006, images of water flowing to ocean in times of drought strengthened the public perception of stormwater as a valuable resource.
National water reform agenda	Anticipating to forecasted climate change, population growth and urban development, the Government of Australia has set a framework for urban water reform in 2004 through the National Water Initiative (NWI) that stimulate a series of key water supply, efficiency and pricing innovations to achieve reliable, healthy, safe and sustainable urban water supply and more liveable, sustainable and economically prosperous cities. Amongst others, the NWI has developed principles for urban water planning that stimulate a holistic water cycle approach and consideration of the full portfolio for water supply and demand options (see National Water Commission, 2009, 2011).

such, this paradigm shift towards at-source stormwater treatment was instrumental for the adoption of stormwater as an alternative water resource. Moreover, the network of advocates who pushed for innovative stormwater treatment supported the uptake of stormwater harvesting and reuse through their stormwater management expertise and their access to relevant actors. This network of WSUD advocates was concerned with the dominant focus of the state governments towards centralised augmentation of water supplies through desalination and argued that a combination of centralised and decentralised water resources (i.e. recycled wastewater and greywater and harvested rainwater and stormwater) and efficient water consumption would be more sustainable in the long term.

The WSUD advocates adopted a strategy of network building, experimentation, research and capacity building to stimulate the uptake of stormwater harvesting and reuse. Networks were built around real projects in which innovative stormwater schemes were implemented in practice, mostly driven by the need to overcome hurdles in the planning and realisation processes (e.g. conflicting or absent regulation, limited cross-disciplinary knowledge). In addition, research projects such as the Cities as Water Supply Catchments research program provided a platform for collaboration between the key network actors (including public and private sectors and academia).

As such, the uptake of stormwater harvesting and reuse evolved through a process that was driven by a learning and advocacy agenda rather than by deliberate public policy development.

6.4.2.3 Governance of delivery of stormwater harvesting and reuse

As the involved individuals have no or limited previous experience with the planning, realisation, operation and maintenance of the technology, stormwater harvesting and reuse projects have an experimental character. With regards to implementation, there is limited coordination taking place to exchange experiences and resources between stormwater harvesting and reuse projects. Interaction between projects is rather taking place ad hoc, mostly through informal networks of project managers at various involved organisations to exchange experiences and best practices for overcoming shared issues.

However, there are several examples of coordination that contradict the overall perception of the interviewed individuals. For example, in Adelaide a coordinated approach is taken to implement eight stormwater harvesting and reuse schemes that are co-funded through the aforementioned National Urban Water and Desalination Plan. As a consequence, resources are coordinated and experiences are exchanged amongst these projects. Moreover, the projects have fed into overarching urban water policies, Water for Good (2010) and the Stormwater Strategy (2011), so that experiences are adopted by strategic planning for urban water management in the future. Other examples of coordination support the implementation of stormwater harvesting through knowledge development, capacity building and community building. For example, in Melbourne the Clearwater programme facilitates the network of local practitioners through capacity building initiatives and network events that are specifically tailored to enhance the uptake of innovative stormwater technologies.

6.4.2.4 Interaction between strategic planning and adaptation action

In contrast with the case of river widening in the Netherlands, the distinction between governance for strategic planning and delivery of stormwater harvesting is not clear. The adaptation process is being driven by an advocacy and learning agenda and implementation occurs through projects that are being perceived as experiments by the people who undertake them (Table 6.6).

Although the adaptation process is deliberately planned by advocates for stormwater harvesting and their networks and becoming larger and more influential, the concept of stormwater harvesting and reuse is, except for Adelaide, not part of overall metropolitan water policy frameworks across Australia. Strategic planning activities therefore focus on overcoming barriers for implementation of individual projects and building cross-disciplinary professional networks for knowledge exchange between the involved disciplines and organisations. By doing so, these activities have stimulated implementation through enabling prompt responses to make use of newly available national funding opportunities. Furthermore, they have contributed to deliberate learning from the stormwater harvesting projects. The experimental nature of the projects in which stormwater harvesting and reuse has been realised is reinforced through the involvement of research activities that provide scientific knowledge and a platform for strategic partnerships that inform strategic planning (i.e. policy and regulation) for stormwater harvesting and reuse.

Table 6.6. Summary of governance for enhancing the uptake of stormwater harvesting and reuse

	Governance for strategic planning for stormwater harvesting and reuse	Governance of delivery of stormwater harvesting and reuse
Timing	Ongoing since the early 1990s, with an experimental character which intensified since the peak of the drought in 2006.	Idem, mostly focused around demonstration projects.
Relationships	Cross-disciplinary network building between policy, practice and science for reasons of knowledge exchange, overcoming hurdles for implementation and advocacy.	Idem.
Knowledge	Analysis of how individual projects can support a vision of WSUD.	Deliberate collaborative learning for implementation of 'experimental' stormwater harvesting and reuse projects.
Financial resources	Allocation of a $200 million national fund for stormwater harvesting and reuse projects.	Acquisition of national fund and co-funding arrangements; financial reporting.

6.5 Comparative analysis

In Section 6.3, it is hypothesised that adaptation processes are cyclic processes in which strategic planning for adaptation and the delivery of adaptation are influencing each other (see also Figure 6.1). Both cases confirm that the strategic planning and delivery sides of governance of adaptation do indeed mutually reinforce each other: governance for strategic planning of adaptation creates legitimacy and effective adaptation action; whilst governance of delivery of adaptation influences future strategic planning through new relationships and knowledge that have been established during the implementation of adaptation action.

The case of river widening in the Netherlands illustrates that strategic planning for adaptation can effectively lead to system wide implementation of adaptation action. More specifically, the case shows how a systems approach towards adaptation and the organisation thereof leads to a coherent programme vision with a legitimate set of adaptation measures that are delivered by means of a deliberate learning agenda to achieve the programme's objectives. Similar to the theory of (Table 6.1), governance for strategic planning of river widening focused on engaging stakeholders for the development of a legitimate, feasible and realistic set of measures and creating the boundary conditions (i.e. budget, objectives, roles) for the delivery of the measures, whilst governance of delivery of the measures focused on task setting, monitoring, facilitation, community building, anticipating and responding to emerging issues (Table 6.4). Governance of delivery of the programme has been influencing future strategic planning processes through facilitation of individual projects (e.g. impact on policy and new tools and guidelines) and building communities building amongst the politicians and professionals involved. As such, it can be concluded that governance for strategic planning and governance of delivery of the river widening measures are mutually reinforcing in this particular case.

The case of stormwater harvesting and reuse in Australia illustrates that, in contrast to the river widening case, the distinction between governance for strategic planning and governance of delivery of adaptation is not always clear. The aim of both elements of governance of adaptation overlaps because of the experimental character of the delivery of adaptation action: experimental projects are realised to learn about the feasibility and effectiveness of the stormwater harvesting and reuse technology. However, despite the difficulty to make a clear distinction between governance for strategic planning and governance of delivery of adaptation action in the stormwater case, most elements of Table 6.1 were identified in the stormwater case (Table 6.6). From the perspective of strategic planning, these experimental projects play an important role in exploring the available adaptation options and establish new relationships (and supporting institutional arrangements) to realise such new options. From the perspective of delivery of these projects, collaborative learning plays an important role to develop the schemes and overcome challenges related to the planning, design, regulation, operation and maintenance. As these aspects cannot be considered in isolation of each other, it can be concluded that governance for strategic planning and governance of delivery of the stormwater harvesting and reuse schemes are mutually reinforcing in respect to realising stormwater harvesting and reuse schemes.

It should be noted that, in contrast to the theoretical framing of Table 6.1, the dominant focus of the strategic planning for these 'experimental' stormwater projects is mainly directed at the enabling and constraining factors of individual projects rather than improving the functionality of urban water systems as a whole. However, stormwater harvesting and reuse has emerged as an alternative water resource that is now considered to play an important role in the overarching vision for water sensitive cities which focuses on the functionality of the urban water system as a whole.

Table 6.7 Interactions between governance for strategic planning and delivery of adaptation

	From strategic planning to delivery	**From delivery to strategic planning**
Relationships	Stakeholder support and/or commitment for the realisation of adaptation action (NL, AUS).	Matured relationships with increased level of trust as a result of collaboration to jointly achieve projects (NL, AUS).
Knowledge	A broad context-specific knowledge base (technical, socio-political, institutional) for the implementation of projects (NL, AUS).	New/adapted policy, legislation and regulation based on experience about overcoming practical challenges in projects (NL, AUS).
Financial resources	Allocation of budgets to invest in adaptation (NL, AUS)	Reconsidering and sometimes cancelling projects after over-achieving projects within a programme (NL).

Based on the presented findings of the two cases, Table 6.7 summarises the interactions through which governance for strategic planning and delivery of adaptation action are mutually reinforcing. Governance of strategic planning is effective when the conditions are created that are needed to deliver adaptation action effectively, including stakeholder support, a broad knowledge base and an allocated investment budget for the realisation of adaptation action. Furthermore, the knowledge and relationships that are developed for the realisation of adaptation action can be beneficial for strategic planning of new adaptation actions. However, both cases demonstrate that the connection from delivery to strategic planning is not straightforward. As a consequence of the 'experimental' character, planning processes for realising stormwater harvesting and reuse schemes need to be gone through repeatedly and any upcoming regulatory barriers need to be tackled on a scale of individual projects. This enables the establishment of tailor-made designs, but also requires repeated action to overcome hurdles in the planning process. In comparison, regulatory hurdles in the planning processes for river widening in the Netherlands were, if possible, simultaneously tackled through a centralised approach for the whole Room for the River programme. However, this has demanded significant preparation during the initiation phase of the programme (2000-2006).

6.6 Conclusion: advancing adaptation action

Governance of adaptation is not only about strategic policy making, but also about delivery of adaptation. The cases of river widening in the Netherlands and stormwater harvesting and reuse in Australia illustrate that governance for strategic planning of adaptation and governance for delivery of adaptation are mutually reinforcing. Governance of strategic planning has the ability to reinforce delivery through creating the conditions that are needed to deliver adaptation action effectively, including stakeholder support, a broad knowledge base and an allocated investment budget for the realisation of adaptation action. Conversely, governance of delivery can be influential for strategic planning of new adaptation actions through knowledge and relationships that are developed for the realisation of adaptation action. Because governance for strategic planning of adaptation and governance for delivery of adaptation are mutually reinforcing, they should not be considered in isolation of each other. In other words, during strategic planning processes, the governance for delivery should be taken into account in order to effectively realise adaptation action. And vice versa, during the delivery of such action, attention should be paid to the implications for strategic planning of future adaptation. Thus far, this distinction is insufficiently acknowledged in the literature about governance of adaptation.

As outlined in Sections 6.1 and 6.2, contemporary research on governance of adaptation faces difficulties to address delivery of adaptation effectively. Thus far, the dominant focus of this body of research is on strategic planning for adaptation. Based on the two case studies that are presented here, I suggest that adaptation research should pay more attention to the governance of delivery of adaptation action as this could help overcome governance challenges that impede adaptation action. The river widening case clearly illustrates how a large scale programme can bring together governance for strategic planning and delivery. In addition, the stormwater case shows that a deliberate learning agenda of a professional network of advocates can bring together both elements of governance of adaptation. However, the findings of these two cases are insufficient for general recommendations about how to improve the uptake of adaptation action. Therefore, I suggest that future research further investigates the role of both by continuing to investigate the role of formal and informal networks, as is nowadays increasingly being done (e.g. Bodin and Crona, 2009; Juhola and Westerhoff, 2011; Olsson et al., 2006), and by using lessons from project management to enrich the existing strategic focus of governance of adaptation.

Furthermore, I note that adaptation follows different, but complementary, trajectories in both cases. These trajectories have in common that adaptation action is often catalysed by a crisis (respectively near-miss floods and drought), but the scale and pattern of implementation differ significantly. In the case of river widening in the Netherlands, a programmed approach for adaptation on a system scale was preceded by a prolonged period of envisioning and experimentation from the 1970s onwards. During this period, the added value and feasibility of river widening and combining multiple functions in flood prone areas were explored. As such, this early period has similarities with the largely 'experimental' nature of the uptake of stormwater harvesting and reuse in Australia. The advocates of stormwater harvesting, and others who are involved in experimental stages of innovation, could potentially use insights from the theory on project management (particularly its sub-domains of programme management and process management; e.g. Hertogh and Westerveld, 2010; Rijke et al., in press; Shehu and Akintoye, 2009) to turn these experiments into successful programmes. It should be acknowledged that there are possibly many other adaptation trajectories that are also worthy of investigation. Therefore, there is a need for more empirical testing of the causes of success and failure of delivering adaptation action through alternative governance approaches during different adaptation trajectories.

Based on the analysis, it is clear that the scholarly communities of governance of adaptation and project and programme management need to work more closely together as a way of addressing adaptation action effectively. Hence, I suggest that the governance of adaptation scholarship refocuses its current emphasis on strategic planning ('management as learning') towards a research agenda that also incorporates a lens for implementation in order to turn aspirations into reality ('management as learning for achieving results').

CHAPTER SEVEN

Synthesis

Delivering change.

This chapter combines the insights regarding the 'what', 'when', 'why' and 'how' questions to answer the overarching research question of how to deliver adaptation effectively. The main research findings are synthesised in this chapter and a reflection is made upon the set of research papers that provided the basis for this thesis.

7. Synthesis

7.1 Conclusions, reflections and recommendations

7.1.1 What?

Research question: What is effective governance of adaptation?

Conclusion

As highlighted in chapters 3 and 6, the scientific literature related to governance of adaptation suggests that the effectiveness of governance of adaptation is the combined capability of governance to anticipate a change (adaptive governance) and to establish system change (transition management). In this thesis, I have introduced the concept of fit-for-purpose governance as an indication of the effectiveness of governance structures and processes at a certain point in time for fulfilling the functions of a particular system, such as an urban water system, under a given set of contextual conditions. The fit-for-purpose of governance can be assessed through a procedure that consists of three steps: (1) identifying the purpose of governance; (2) mapping of the context; and (3) evaluating the outcome of governance mechanisms.

Reflection

The concept of fit-for purpose governance was initially developed for making adaptive governance operational (chapter 2). It was introduced as a supplement to the concept of adaptive governance to make explicit the inherent uncertainties that policy makers and practitioners are dealing with: whilst adaptive governance focuses on responding to (potential) change, fit-for-purpose governance is specifically considering the (future) functions that the social and physical components of a particular system, such as an urban water system, have to fulfil. However, in chapter 3 it was shown that the concept of fit-for-purpose governance is applicable for both adaptive governance and transition management as it addresses different stages of transformation. As such, it can be concluded that the concept of fit-for-purpose governance can indeed be applied as an indication for the effectiveness of governance of adaptation for both anticipating and establishing change.

As described in chapter 2, important constraints to the uptake of adaptive governance relate, to a large extent, to the inability of practitioners and policy makers to cope with complexity and uncertainties. In particular embracing complexity and uncertainty, continuous learning, and ongoing reflection and adjustment of management approaches, are providing practical challenges because they are not being institutionalised into planning practice. In contrast, chapter 2 argues that the fit-for-purpose procedure would form a better match with existing institutional frameworks, because it is providing a snapshot of the current situation rather than a prediction of the (uncertain) future. Through intensive stakeholder engagement, the procedure proposes to reduce uncertainties regarding the current situation and strive for 'good enough' governance that is fit for its purpose under given conditions. Although empirical evidence from practitioners' experiences with the procedure is not yet available, the fit-for-purpose governance framework provides the basis for a new way of thinking for overcoming impediments for adaptive governance that is based on existing institutional frameworks of predict and control regimes in which most policy makers operate (see chapter 2). It would, therefore, be worthwhile for scientists interesting in developing guidance to policy and decision makers for implementing adaptive governance to explore whether advantage can indeed be gained from using this perspective.

Chapter 2 concludes with a statement that the concept of fit-for-purpose governance is not yet readily applicable in governance practice, because empirical evidence is lacking to show how the framework works in practice. Furthermore, chapter 2 concludes that further research is needed to operationalise the concept of fit-for-purpose governance. Filling this void, it is illustrated in chapter 3 how the framework can be applied. In chapter 3, the stakeholders' perceptions of the strengths and weaknesses of employed governance approaches are being used as a proxy for the fit as the effectiveness of interactive processes arguably depends on how satisfied stakeholders are with these processes. For the case studies of Australian urban water management, this has enhanced understanding of what governance approaches are effective during different transformation stages (see also the next conclusion in section 7.2.1). In chapter 2, I also suggest that receptiveness and advocacy could also be indicators of fit of governance approaches to their social and physical contexts, as these factors relate to the willingness of actors to accept and invest (e.g. time, effort, capital and/or reputation) in these approaches. The applicability of these proxies remains yet to be tested.

The application of the fit-for-purpose framework (chapter 3) has shown that this is a time consuming process that requires the input of all key stakeholder groups to get a comprehensive overview about the purposes, context and anticipated governance outcomes. Broad stakeholder involvement is needed to take into account multiple perspectives of these aspects. However, at the same time the framework is set up to serve its users to develop governance reforms that are fit to a certain purpose. The framework can therefore only be applied effectively when a purpose is selected explicitly and subsequently all stakeholders reflect on the drivers and barriers to realise that particular purpose in practice (e.g. widespread implementation of stormwater harvesting and reuse schemes to make Australian urban water systems more resilient to drought). The approach that was applied in the research that is presented in chapter 3 first collected the stakeholder perspectives individually and later presented the aggregate findings in workshops in which the representatives of the key stakeholder groups participated in discussions about these findings. Although alternatives may also be appropriate, this approach provided the opportunity for all stakeholders to share their ideas individually and discuss the (urban water) governance context collectively. As such, the process of conducting the fit-for-purpose assessment itself functioned as a platform for developing adequate and legitimate governance reform.

Recommendations

The following recommendations for practice can be derived from the conclusions and reflections above:

- ✓ The application of the procedure for assessing the fit-for-purpose of governance is recommended for policy makers in order to prompt appropriate governance reforms.
- ✓ The fit-for-purpose governance framework should be considered in combination with the elements of adaptive governance that focus on continuous learning and adjustment. It is, therefore, important for policy makers to regularly apply the fit-for-purpose governance procedure in order to adequately anticipate the changing contextual conditions.
- ✓ It is recommended that the fit-for-purpose governance assessment is conducted in a multi-stakeholder setting to capture all relevant perspectives on the purpose, context and anticipated governance out-

comes and provide a platform for adequate and legitimate governance reform.

In addition, the reflection highlighted the following implication for further research:

- ✓ As empirical evidence of practitioners' experiences with the fit-for-purpose governance framework is lacking, it is recommended to analyse whether taking the perspective of practitioners would indeed provide opportunities to overcome impediments to implement adaptive governance in practice.
- ✓ Further research is recommended to improve methods for conducting fit-for-purpose governance assessments. In particular, the applicability of proxies for indicating the effectiveness of governance requires further research.

7.1.2 When?

Research question: When, during different stages of transformation, is a particular governance approach effective?

Conclusion

Based on the insights of chapter 3, it can be concluded that the effectiveness of governance approaches varies during the course of transformation processes. It was shown in chapter 3 that different stages of transformation favour different configurations of centralised/decentralised and formal/informal governance. This insight may be used to develop a pattern of governance approaches that are typically effective during the early, mid and late stages of transformation (see Table 3.5). In summary, chapter 3 suggests that early stages of transformation favour decentralised and informal governance approaches to enable incubation of innovation and formation of informal networks. Mid stages of transformation favour hybrid governance configurations in which formal policy decisions and centralised capacity building efforts are, for example, used to reinforce the decentralised and informal governance of the early transformation stages. Late stages of transformation favour centralised and formal governance to adjust or establish legislative frameworks and coordinate capacity building to convince and enable laggards to adopt innovative approaches and safeguard the reformed practice.

Reflection

The developed pattern of effective governance that is described above provides a first attempt to specify often advocated calls for multi-level governance through a pattern of appropriate multi-level governance configurations during each stage of the process of transformation. The described pattern points out that this does not necessarily mean that centralised approaches need to be overthrown and replaced by decentralised approaches as is sometimes suggested by the literature about adaptive governance (e.g. Nelson et al., 2007; Pahl-Wostl, 2007; van der Brugge and Rotmans, 2007). However, the pattern is specific to the context of urban water governance in Australia and has not been tested in other contexts. Furthermore, the pattern merely took the temporal dimension of system transformation into account and overlooked spatial scales of institutional and infrastructure systems. Comparison with other contexts and different types of infrastructure systems is therefore suggested as needed in order to increase the general applicability of the pattern observed here. Therefore, it remains, in any case, important to identify the purpose, map the context, and assess the anticipated outcomes to customise proposed governance reforms to their contextual conditions.

By putting the stages of the adaptation process and the transition process in parallel, the similarities between the subsequent stages of transition and adaptation processes and the typical activities in these stages have become clearly apparent. Therefore, I refer to transformation rather than adaptation or transition. The similarities between adaptation and transition processes give reason to question whether the social-ecological and socio-technical perspectives on governance are using different words for identical phenomena. Similar approaches, such as multi-level governance, social learning, and (informal) network management are advocated by each of these perspectives. It could be argued that there is a difference, because adaptive governance (social-ecological perspective) is about anticipating change through continuous learning and adjusting, whilst transition management (socio-technical perspective) is about establishing systemic change in the long term. However, this difference may be in practice less apparent than it appears at first sight, because deliberately anticipating change (adaptive governance) requires deliberate planning to establish new practice (transition management). In contrast, it should be noted that the establishment of a new dominant practice (a transition) is in fact triggered typically by changing circumstances and/or insights and occurs through continuous learning and adjustment. A planned transition (as far as possible) is therefore not fundamentally different from planned adaptation. Furthermore, in relation to water management, transition management is at present generally aiming to increase the resilience of vulnerable systems to crises such as floods and droughts. Transition management is therefore aiming *inter alia* to enhance adaptive governance.

It is therefore not surprising that both of the theoretical perspectives (social-ecological and socio-technical) on governance are increasingly converging. Several comparative studies have suggested the need to combine the elements of each approach, for example to create synergy around the topics of goal-setting, stakeholder collaboration, addressing spatial and temporal scales and analysing governance processes for delivering change (e.g. Foxon et al., 2009; Smith and Stirling, 2010; van der Brugge and van Raak, 2007). In addition, these potential cross-connections between the two perspectives were supported amongst scholars involved in both domains during extensive discussions during the second and third International Conference on Sustainability Transitions that were held in Lund, Sweden (13-15 June, 2011) and Copenhagen, Denmark (29-31 August, 2012).

Recommendations

The following recommendation for practice can be derived from the conclusion and reflection above:

- ✓ Based on the pattern of effective governance approaches in the Australian urban water sector, it is recommended to rethink and, if needed, reform the applied governance approaches during the various stages of transformation.

In addition, the reflection highlighted several implications for further research:

- ✓ Further research is recommended to test the pattern in contexts other than the Australian urban water sector and for infrastructure systems of different spatial scales in order to improve its applicability for policy makers to enhance system transformation.
- ✓ Based on the similarities between transition and adaptation processes and the way these processes are managed, albeit for different reasons, it is recommended to further investigate the opportunities to mutually reinforce the socio-technical and social-ecological perspectives on governance of adaptation.

7.1.3 Why?

Research question: Why are transformational processes sometimes being hampered?

Conclusion

Based on the insights of chapter 4, it can be concluded that system transformation depends on the presence of eight enabling factors. These factors include: (1) a narrative, metaphor and image that support a clear vision for change; (2) a regulatory and compliance agenda; (3) economic justification; (4) policy and planning frameworks and institutional design; (5) leadership; (6) capacity building and demonstration; (7) public engagement and behaviour change; (8) research and partnerships with policy/practice. Factors (1) to (4) are requirements for developing and performing new practices, whilst factors (5) to (8) are needed for enabling new practices. Insight into the absence of one or more of the eight enabling factors for system transformation gives insight as to what governance arrangements are not fit for the purpose of delivering change and, thus, why such transformational processes are hampered by inadequate governance.

Reflection

The enabling factors that are described above have emerged from a series of studies in the Australian urban water sector (Farrelly et al., 2012; see Appendix A), where they were subsequently applied to describe governance reforms directed towards water sensitive cities (Rijke et al., 2012b; see Appendix B). The application of the factors in the context of Dutch river flood risk management (chapter 4) has validated the eight factors, but has also revealed that there are other factors required for successfully establishing transformational change. Chapter 4 highlights, for example, that crises such as near-miss floods (or drought, see chapters 3 and 6) can contribute significantly to transformational processes when they act as a catalyst for policy reform. However, it is not possible (nor ethical) to deliberately plan for crises caused by natural phenomena (e.g. Simm, 2012). The set of eight factors should therefore be considered as a checklist of the availability of the required ingredients for change. As such, it enriches other diagnostic tools for governance of adaptation, which often focus on the application of principles (e.g. management as learning, learning by doing, involvement of multiple stakeholders, robustness and flexibility of processes) rather than elements (e.g. Huntjens et al., 2012; Pahl-Wostl et al., 2010).

However, the eighth enabling factor about research partnerships requires nuance. The initial framework, which is based upon the Australian urban water context (see Appendix A), suggests that long-lasting research partnerships with policy and practice are required for the purpose of collaborative learning to establish systemic transformation. Whilst collaborative learning amongst stakeholders played a significant role in the realisation of the Room for the River programme (see van Herk et al., 2013), long-lasting partnerships with research did not take place for purposes of creating systemic transformation. The only structural involvement of researchers and independent experts occurred through the so-called 'Q-team' that was set up to assess and advise about the spatial quality of the individual plans in the programme. The involvement of researchers in the Dutch context of Room for the River has more a character of monitoring the plans, designs and processes by researchers compared to the Australian urban water context where research partnerships between deliverers and researchers are a platform for stakeholder engagement for collaborative learning and knowledge exchange, innovation through generating new ideas and removing barriers for implementation and advocacy to influence policy makers. An explanation for this disparity may lie in the different characteristics of governance of adaptation during the transformation processes in each of the two cases (chapter 6): the experimental character of the delivery of adaptation in Australian urban water systems requires scientific input; whereas this is less apparent in the programmed character of the Dutch case.

Recommendations

The following recommendations for practice can be derived from the conclusion and reflection above:

- ✓ The set of eight enabling factors should be considered as a checklist of the availability of the required ingredients for establishing systemic transformation of water systems. If such transformation is being hampered, the set of enabling factors directs its users to what governance arrangements may be lacking or underdeveloped for establishing systemic transformation and, thus, why transformation is being hampered.
- ✓ As the set of eight enabling factors only provides a snapshot insight into the availability of the required ingredients for change, it is recommended that it be used iteratively in combination with the fit-for-purpose framework to evaluate the effectiveness of the governance structures and processes that are in place.

In addition, the reflection highlighted the following implication for further research:

- ✓ Application and testing of the set of eight enabling factors is recommended to improve its value and use for contexts with transformation pathways that differ from the studied pathways in Australia and the Netherlands.

7.1.4 How?

Research question: How can defined adaptation projects be realised effectively?

Conclusion

The Room of the River case study was used to analyse how adaptation projects can be realised effectively (chapter 5). Room for the River is a programme that aims to increase the level of flood protection through 39 individual river widening projects. The analysis of the Room for the River case study revealed that a combined strategic/performance focus at the level of both programme and project management can lead to a collaborative approach between programme and project management. This enables effective stakeholder collaboration to enhance the legitimacy and quality of the programme and its projects and adaptation of the programme's organisation to contextual changes, newly acquired insights and the changing needs of consecutive planning stages. The case study showed that these two factors contributed significantly to establishing five presumed success factors for effective programme management: (1) a clear programme vision that is widely supported by all relevant stakeholders; (2) a clear priority focus that provides opportunities to connect stakeholder ambitions to the overall programme objectives; (3) a transparent programme planning framework that outlines the boundary conditions and roles of the stakeholders; (4) an appropriate programme strategy that matches the vision, priority focus and planning framework of the programme. In case of Room for the River a strategy of decentralised decision making within centralised boundaries was followed; (5) appropriate programme coordination to monitor progress and performance and assist projects in achieving their objectives.

Reflection

With Room for the River as a case study, this thesis has focused on the realisation of adaptation through a 'programmed approach' in which individual adaptation projects are part of an overarching programme. Programmed approaches to adaptation are increasingly being applied to develop adaptation policies and plans (e.g. the Delta Programme in the Netherlands and the Californian Statewide Flood Management Planning Program in the United States) and implement adaptation measures (e.g. the national flood defence programmes in the Netherlands; HWBP, HWBP-2 and nHWBP). The nature and context of these programmes can vary significantly, which requires customised programme management approaches. The insights from the analysis of the Room for the River programme should, therefore, be considered as a source of inspiration rather than a blueprint for effective programme management. Similarly, the insights from the Room for the River case study may provide valuable elements for effectively realising stand-alone projects that are not part of a programme. However, the interactions between project and programme management do obviously not apply to stand-alone projects.

The case of the Room for the River programme illustrates that the adaptive approach to the governance of water resources contains relevant elements for the management of programmes and projects. Adaptive governance of water resource systems and adaptive management of programmes and projects have in common that they aim to anticipate a change, involve multiple actors and sources of knowledge to make balanced decisions, and make adjustments based on monitoring and learning. However, both are being addressed by two separate bodies of literature that exist largely in isolation from each other. It is therefore logical to further explore the transferability between both bodies of literature. As adaptive governance and adaptive project management differ in terms of scope and time horizons (chapter 6), this could potentially generate new insights. For example, a concept such as continuous learning from the adaptive governance literature could be relevant to project management scholarship which has recognised the challenge to maintain knowledge after projects have been completed. In contrast, adaptive governance could benefit from concepts from the project management literature, such as 'iterative life cycle management' and 'agile management' (Project Management Institute, 2008a), in which actions are monitored and, if needed, adjusted with fixed iterations in terms of time (in the order of weeks/months). Whilst this thesis provides a first attempt to bring together both bodies of literature, I suggest it is necessary to further explore the cross-connections in more detail.

Recommendations

The following recommendation for practice can be derived from the conclusion and reflection above:

- ✓ Application of the identified attributes for effective programme management is suggested for policy makers, programme architects and programme managers to set in place required elements (vision, priority focus, planning framework, strategy, coordination mechanisms) and adequate processes (stakeholder engagement, programme adaptation) for successful delivery of planned adaptation action. However, it is important to customise the programme management organisation to the nature and context of the programme.

In addition, the reflection highlighted several implications for further research:

- ✓ Further research is recommended to analyse the applicability of the attributes for effective programme management for other programme management typologies, such as portfolio management and programmes as service centers, and other sectors.

- ✓ As adaptation action is often being delivered through projects and programmes, it is recommended to researchers in the domain of governance of adaptation to investigate the relevance and effectiveness of project and programme management approaches for the governance of adaptation.
- ✓ Further research is recommended to investigate the applicability of adaptive management approaches for projects and programmes and their effectiveness for realising project/programme objectives within their boundary conditions.

Research question: How can strategic planning enhance the implementation of adaptation action effectively?

Conclusion

Based on the insights of chapter 6, it can be concluded that governance for strategic planning and also for the delivery of adaptation have the ability to mutually reinforce each other. Governance of strategic planning has the ability to enhance delivery through creating the conditions that are needed to deliver adaptation action effectively, including stakeholder support, a broad knowledge base and an allocated investment budget for the realisation of adaptation action. In parallel, governance of delivery can be influential for strategic planning of new adaptation actions through knowledge and relationships that are developed for the realisation of adaptation action. The cases of Room for the River (chapter 4 and 6) and Australian urban water management (chapter 3 and 6) illustrate that respectively large scale infrastructure programmes and informal professional networks can play an important role in connecting strategic planning for adaptation and delivery of adaptation action in practice.

Reflection

As outlined in chapter 6, governance of adaptation is about strategic planning *and* the delivery of adaptation. The notion that governance for strategic planning and delivery of adaptation are interdependent is relevant to the practice and science of governance of adaptation. With regard to practice, governance of strategic planning of adaptation and governance of the delivery of adaptation should take account of each other. During strategic planning processes, the governance for delivery should be taken into account in order to develop effective adaptation actions that can be realised effectively. And, during the delivery of adaptation action, attention should be paid to the implications for strategic planning of future adaptation. The findings of chapter 6 illustrate that such an approach to the governance of adaptation can overcome barriers to adaptation. However, as highlighted in chapters 4 and 6, it is not self-evident in practice that lessons from the delivery of adaptation are being adopted in governance processes for the strategic planning of consecutive adaptation actions.

In order to take effective adaptation action and avoid under or overinvestment in adaptation it is important to take into account opportunities to link adaptation measures to developments in other domains, such as urban development, agriculture or business, for achieving synergy in terms of cost-effectiveness, reduced hindrance or higher quality (e.g. Gersonius et al., 2012; Veerbeek et al., 2012). This requires shifting between (spatial, temporal and institutional) scales as opportunities to link adaptation measures often emerge at short term and local scales and sometimes involve different stakeholders than the scales on which strategic planning processes typically focus. The implication of this notion for the governance of adaptation is that a different way of policy making is needed, because traditional ways of technocratic policy cycles - in which acknowledgement of a problem consecutively leads to formulation of new policy, development of a solution and maintenance and operation of that solution (e.g. May and Wildavsky, 1978; Winsemius, 1986) – are no longer valid as they cannot take into account local opportunities that may arise during the planning and realisation of specific projects. However, the findings of chapter 6 suggest that the governance of delivery of adaptation is insufficiently acknowledged by the literature dealing with governance of adaptation. It is, therefore, interesting to the research domain related to the governance of adaptation to further explore the relationship between governance of strategic planning and the governance of delivery of adaptation, because this could help to overcome governance challenges that impede adaptation action.

The findings of chapter 6 illustrate that programmes and informal networks could play an important role in bridging policy and practice, because they typically involve key stakeholders that are involved in strategic planning and delivery of adaptation and aim at influencing existing policies to realise their own objectives and ambitions. Further research towards bridging both elements of governance of adaptation is therefore recommended. The role of informal networks in relation to governance of adaptation is nowadays being increasingly studied. Nevertheless, it remains a challenge to measure and provide guidance for increasing the effectiveness of informal networks for influencing adaptation action. In contrast to the attention to informal networks, the role of programmes for bridging governance of strategic planning and delivery of adaptation is not yet a common research topic. Projects with an experimental nature are often the subject of research, particularly in relation to social learning, but large scale projects and large scale investment programmes less frequently studied in the context of adapting infrastructural systems. This is particularly remarkable because such projects and programmes arguably contribute more to the magnitude and speed of adaptation compared to demonstration projects (see chapter 6). Hence, it would be worthwhile to investigate the role and influence of large scale projects and programmes related to the governance of adaptation.

Recommendations

The following recommendations for practice can be derived from the conclusion and reflection above:

- ✓ Policy makers and politicians are advised to take account of the governance of the delivery of adaptation during strategic planning processes for adaptation. Meanwhile, project managers, politicians and other stakeholders involved in adaptation projects are urged to pay attention to the implications of their role and actions for assisting strategic planning of future adaptation.
- ✓ A different approach to policy making is recommended that combines a policy driven approach (driven by a certain policy objective) with an opportunistic approach (driven by local opportunities to connect investment agendas) in order to adapt effectively.

In addition, the reflection highlighted several implications for further research:

- ✓ Further research investigating the relationship between governance of strategic planning and governance of delivery of adaptation is recommended. Traditional thinking about policy cycles needs to be reassessed for adaptation in order to incorporate opportunities and challenges that emerge during the implementation of policies.
- ✓ Further research is recommended regarding the role of informal networks for connecting strategic planning and delivery of adaptation and specifically for enhancing the capacity to effectively deliver adaptation action.
- ✓ Because the Room for the River case suggests that large scale projects can have significant impact on strategic planning for future adaptation actions and follow-up projects, further research is recommended to investigate the role and influence of large scale projects and programmes on the governance of adaptation. Understanding of how such projects inform policy making, set precedents for future practice and how the lessons of these programmes can be embedded within the involved organisations could be used to increase the effectiveness of the governance of adaptation.

7.2 Concluding reflection

The main research question of this thesis is:

How can adaptation actions to manage changes in flood and drought risks be delivered effectively?

By focusing on the questions of what, when, why and how, this thesis provides several ingredients for addressing the question (section 7.1). These include: a procedure for establishing fit-for-purpose governance reform (WHAT; chapter 2); a pattern of governance approaches that are typically effective during the early, mid and late stages of transformation (WHEN; chapter 3); a checklist for the availability of the required ingredients for change (WHY; chapter 4); guidance for effective design and management of adaptation programmes (HOW; chapter 5), and recommendations for aligning governance of strategic planning and delivery of adaptation (HOW; chapter 6). These ingredients provide principles and attributes that contribute to the effective delivery of adaptation to flooding and drought, which need to be customised to the context and purpose of their users in practice. However, as pointed out in the previous section, they also lead to a number of topics for further research, particularly because the findings that are presented in this thesis are based on a limited number of cases.

I am ending this thesis with several overall reflections. Firstly, this thesis does only implicitly address the question of who is involved and should be involved with the governance of adaptation. Governance is a purposeful action with a purpose that is subject to stakeholder perspectives (chapter 2). As a consequence of the demand driven approach of this research, this thesis took the perspective of the advocates for stormwater harvesting and reuse in the Australian urban water context and the advocates for river widening for flood risk management in the Netherlands. With these perspectives arbitrarily chosen, less emphasis was given to the question of who governs and who should govern adaptation. With the shift from 'government' to 'governance' (chapter 2), decision making about water management is no longer a matter that is exclusive to the public sector. Instead, a wide range of actors, such as the private sector, academia and citizens, are involved with the planning, financing, realisation, operation and maintenance of water infrastructure. As repeatedly recalled throughout this thesis, panacea, blueprint solutions, for effective water governance do not exist and governance approaches should be specifically developed, case-by-case. Hence, stakeholder participation should also be customised to the nature of policies, plans or projects and their context. As such, this thesis should be considered in combination with the work of others who have made more detailed analyses of the various and many approaches to stakeholder participation (e.g. Edelenbos and Klijn, 2006; Edelenbos et al., 2011; Reed, 2008).

Secondly, this thesis uses case studies with a focus on adaptation to flooding and drought; two hazards that are very different. Obviously, flooding relates to too much water, whilst drought relates to too little availability of water. In addition, flood events, for example, mostly occur rapidly and often have a duration of days to weeks, whereas droughts are persistent in nature and generally have a longer duration than floods (the Australian Millennium drought persevered for nearly a decade). Furthermore, the technological systems that aim to provide flood protection and water security differ significantly in terms of scale, operation and management (chapter 6). Despite these differences, the governance of adaptation to flooding and drought in respectively the Netherlands and Australia share remarkable similarities. Chapter 6, describes, for example, how a crisis (respectively near-miss flooding and the peak of the drought) first triggered adaptation through infrastructure upgrades and after which governance reform followed. Although both cases followed different adaptation trajectories (i.e. programmed vs. experimental), they both revealed that establishing the required connections between organisations, departments, disciplines, domains, sectors, management levels and individuals for realising adaptive systems is crucial, but challenging. The similarities that the two cases reveal for governance of adaptation to flooding and drought suggest that the lessons and experiences from adaptation to drought can be used as an inspiration to adapt to flooding and vice versa. This may be relevant for water management contexts that are vulnerable to both flooding and drought, such as Australian cities.

Thirdly, during the PhD research I have met many colleague scientists, policy makers and planners who perceive resilience as a virtue. In their view, the purpose of the analysed governance approaches was to establish water systems that are resilient to change; i.e. water systems that have the capacity to absorb shocks whilst maintaining function, and to recover and re-organise after a shock has taken place (chapter 1). This may seem self-evident from an engineering perspective, but from a governance perspective the virtue of resilience requires nuance. Formal institutions, such as legislation, planning authorities and regulators are often set up to serve communities and protect public values, such as public health, the environment and welfare. In this regard, resilience of these institutions provides the capacity to safeguard such values, for example by prohibiting the application of technological systems that cause unacceptable risks for public health or the safety of communities. However, this thesis illustrates that the same formal institutions sometimes hamper adaptation of technological systems and the use thereof when the risks are changing. Furthermore, a fact-finding mission about urban flood management in Saint-Louis, Senegal, in which I participated on behalf of UNESCO-IHE, taught me that, in some contexts, communities have a great ability to cope with flooding, recover quickly and resume life as they knew it. In the case of Saint-Louis, this held particularly true for communities in informal settlements in the floodplain of the River Senegal. Whilst the capacity of these communities to cope and recover from flooding could be classified as resilience; it is probable that these communities would prefer better flood protection over resilience. Hence, it would be worthwhile to systemically investigate the pros and cons of resilience in this type of context and determine when it is a virtue and when it is not. However, this was unfortunately beyond this thesis.

Driven by my personal motivations and academic background, this thesis contributes to the practice and science of governance of adaptation through a demand driven approach in which problem framing and solving were equally important. The research questions that are being addressed in this thesis are more derived from practice than theory. This thesis contains elements that enrich specific research domains, such as the pattern of effective governance during different stages of transformation that enriches the adaptive governance and transition management literature (chapter 3) or the attributes for effective programme management which enriches the programme management literature (chapter 5). However, in my opinion, the greatest implication of this thesis for the research related to the governance of adaptation lies in the notion that science can develop valuable contributions for practice through combining insights from diverse research domains. Based on the analyses that are presented in this thesis, it is therefore likely that the practice of governance of adaptation would benefit from further exploring combinations of disparate literature domains, such as adaptive governance / transition management and adaptive governance / adaptive (project) management.

REFERENCES

References

Abrahamson, E., Rosenkopf, L. (1997) Social network effects on the extent of innovation diffusion: A computer simulation. Organization Science 8, 289-309.

Acheson, J.M. (2006) Institutional failure in resource management. Annu. Rev. Anthropol. 35, 117-134.

Adger, W. (2001) Scales of governance and environmental justice for adaptation and mitigation of climate change. Journal of International Development 13, 921-931.

Adger, W., Dessai, S., Goulden, M., Hulme, M., Lorenzoni, I., Nelson, D., Naess, L., Wolf, J., Wreford, A. (2009) Are there social limits to adaptation to climate change? Climatic Change 93, 335-354.

Adger, W., Hughes, T., Folke, C., Carpenter, S., Rockstrom, J. (2005a) Social-ecological resilience to coastal disasters. Science 309, 1036.

Adger, W.N., Arnell, N.W., Tompkins, E.L. (2005b) Successful adaptation to climate change across scales. Global Environmental Change Part A 15, 77-86.

Adger, W.N., Brown, K., Fairbrass, J., Jordan, A., Paavola, J., Rosendo, S., Seyfang, G. (2003) Governance for sustainability: towards a 'thick' analysis of environmental decision making. Environment and Planning A 35, 1095-1110.

Adger, W.N., Brown, K., Nelson, D.R., Berkes, F., Eakin, H., Folke, C., Galvin, K., Gunderson, L., Goulden, M., O'Brien, K. (2011) Resilience implications of policy responses to climate change. Wiley Interdisciplinary Reviews: Climate Change.

Agrawal, A. (2003) Sustainable Governance of Common-Pool Resources: Context, Methods, and Politics. Annual Review of Anthropology 32, 243-262.

Anema, K., Rijke, J., (2011) Putting new climate adaptation measures into practice: why bother?, Resilient Cities 2011 - 2nd World congress on cities and adaptation to climate change, Bonn, Germany.

Aritua, B., Smith, N.J., Bower, D. (2009) Construction client multi-projects – A complex adaptive systems perspective. International Journal of Project Management 27, 72-79.

Armitage, D., Marschke, M., Plummer, R. (2008) Adaptive co-management and the paradox of learning. Global Environmental Change 18, 86-98.

Armitage, D.R., Berkes, F., Doubleday, N. (2007) Adaptive co-management: collaboration, learning, and multi-level governance. UBC Press, Vancouver, BC, Canada.

Arts, B., Leroy, P., Van Tatenhove, J. (2006) Political modernisation and policy arrangements: a framework for understanding environmental policy change. Public Organization Review 6, 93-106.
Artto, K., Martinsuo, M., Gemünden, H.G., Murtoaro, J. (2009) Foundations of program management: A bibliometric view. International Journal of Project Management 27, 1-18.
Australian Government, (2013) National Urban Water Desalianation Plan - Projects sorted by state, http://www.environment.gov.au/water/policy-programs/urban-water-desalination/.
Avolio, B., Walumbwa, F., Weber, T. (2009) Leadership: Current theories, research, and future directions. Annual Review of Psychology 60, 421-449.
Axelrod, R. (1997) The complexity of cooperation: agent-based models of competition and collaboration. Princeton University Press.
Bass, B. (1985) Leadership and performance beyond expectations. Free Press New York.
Bass, B. (1999) Two decades of research and development in transformational leadership. European Journal of Work and Organizational Psychology 8, 9-32.
Baumgartner, F.R., Jones, B.D. (1991) Agenda Dynamics and Policy Subsystems. The Journal of Politics 53, 1044-1074.
Bazire, M., Brézillon, P., (2005) Understanding Context Before Using It, in: Dey, A., Kokinov, B., Leake, D., Turner, R. (Eds.), Modeling and Using Context. Springer Berlin / Heidelberg, pp. 113-192.
Beringer, C., Jonas, D., Kock, A. (2013) Behavior of internal stakeholders in project portfolio management and its impact on success. International Journal of Project Management.
Berkes, F., (2002) Cross-scale institutional linkages: perspectives from the bottom up, in: Ostrom, E., Dietz, T., Dolsak, N., Stern, P., Stonich, S., Weber, E. (Eds.), The drama of the commons. National Academy Press, Washington, DC, pp. 293-321.
Berkes, F., Folke, C., Colding, J. (2000) Linking social and ecological systems: management practices and social mechanisms for building resilience. Cambridge Univ Pr.
Berkhout, F., Smith, A., Stirling, A. (2004) Socio-technological regimes and transition contexts. System innovation and the transition to sustainability: theory, evidence and policy, 48-75.
Berrang-Ford, L., Ford, J.D., Paterson, J. (2011) Are we adapting to climate change? Global Environmental Change 21, 25-33.
Biesbroek, G.R., Klostermann, J.E., Termeer, C.J., Kabat, P. (2013) On the nature of barriers to climate change adaptation. Regional Environmental Change, 1-11.

Biswas, A.K. (2004) Integrated water resources management: a reassessment. Water international 29, 248-256.
Blomquist, T., Müller, R. (2006) Practices, roles, and responsibilities of middle managers in program and portfolio management. Project Management Journal 37, 52-66.
Boal, K., Schultz, P. (2007) Storytelling, time, and evolution: The role of strategic leadership in complex adaptive systems. The Leadership Quarterly 18, 411-428.
Bodin, Ö., Crona, B. (2009) The role of social networks in natural resource governance: What relational patterns make a difference? Global Environmental Change 19, 366-374.
Bodin, Ö., Crona, B., Ernstson, H. (2006) Social networks in natural resource management: What is there to learn from a structural perspective. Ecology and Society 11, r2.
Bodin, Ö., Norberg, J. (2005) Information network topologies for enhanced local adaptive management. Environmental Management 35, 175-193.
Boisot, M., McKelvey, B. (2010) Integrating modernist and postmodernist perspectives on organizations: A complexity science bridge. Academy of Management Review 35, 415-433.
Bos, J., Brown, R. (2012) Governance experimentation and factors of success in socio-technical transitions in the urban water sector. Technological Forecasting and Social Change 79, 1340–1353.
Bressers, H., Huitema, D., Kuks, S. (1994) Policy networks in Dutch water policy. Environmental politics 3, 24-51.
Bressers, H., O'Toole Jr, L. (1998) The selection of policy instruments: a network-based perspective. Journal of Public Policy 18, 213-239.
Brown, J.T. (2008a) The Handbook of Program Management: How to facilitate project success with optimal program management. McGraw Hill, New York, NY, USA.
Brown, R., Ashley, R., Farrelly, M. (2011) Political and Professional Agency Entrapment: An Agenda for Urban Water Research. Water Resources Management, 1-14.
Brown, R., Clarke, J. (2007) Transition to water sensitive urban design: The story of Melbourne, Australia. Facility for Advancing Water Biofiltration, Monash University.
Brown, R., Farrelly, M. (2009a) Challenges ahead: social and institutional factors influencing sustainable urban stormwater management in Australia. Water science and technology: a journal of the International Association on Water Pollution Research 59, 653.
Brown, R., Farrelly, M. (2009b) Delivering sustainable urban water management: a review of the hurdles we face. Water science and

technology: a journal of the International Association on Water Pollution Research 59, 839.
Brown, R., Farrelly, M., Keath, N. (2009a) Practitioner Perceptions of Social and Institutional Barriers to Advancing a Diverse Water Source Approach in Australia. International Journal of Water Resources Development 25, 15-28.
Brown, R., Keath, N., Wong, T. (2009b) Urban water management in cities: historical, current and future regimes. Water Science & Technology 59, 847-855.
Brown, R.R. (2008b) Local institutional development and organizational change for advancing sustainable urban water futures. Environmental Management 41, 221-233.
Brown, R.R., Farrelly, M.A., Loorbach, D.A. (2013) Actors working the institutions in sustainability transitions: The case of Melbourne's stormwater management. Global Environmental Change 23, 701–718.
Brundtland, G.H., (1987) Report of the World Commission on Environment and Development: Our Common Future. United Nations, World commission on environment and development, New York, USA.
Brunner, R., Steelman, T., Coe-Juell, L., Cromley, C., Edwards, C., Tucker, D., (2005) Adaptive Governance: Integrating Policy, Science, and Decision Making. New York City, NY: Columbia University Press.
Burt, R. (1995) Structural holes: The social structure of competition. Harvard Univ Pr.
Burt, R. (2004) Structural holes and good ideas. American journal of sociology 110, 349-399.
Chen, X., Wang, Z., He, S., Li, F. (2013) Programme management of world bank financed small hydropower development in Zhejiang Province in China. Renewable and Sustainable Energy Reviews 24, 21-31.
Cleland, D.I., Ireland, L.R. (2002) Project management: strategic design and implementation. McGraw-Hill New York.
Commissie Elverding, (2008) Sneller en beter - Advies Commissie Versnelling Besluitvorming Infrastructurele Projecten.
Costanza, R., Daly, H., Folke, C., Hawken, P., Holling, C., McMichael, A., Pimentel, D., Rapport, D. (2000) Managing our environmental portfolio. BioScience 50, 149-155.
Cox, P., Stephenson, D. (2007) A Changing Climate for Prediction. Science 317, 207-208.
Cumming, G.S., Cumming, D.H.M., Redman, C.L. (2006) Scale mismatches in social-ecological systems: causes, consequences, and solutions. Ecology and Society 11, 14.

CW21, (2000) Waterbeleid voor de 21ste eeuw; Geef water de ruimte en aandacht die het verdient. Ministry of Transport and Public Works/Union of Waterboards, The Hague.
Daniels, C., (2010) Adelaide: Water of a City. Wakefield Press, Adelaide, Australia.
Davidson-Hunt, I. (2006) Adaptive learning networks: developing resource management knowledge through social learning forums. Human Ecology 34, 593-614.
Davies, A., Mackenzie, I. (2013) Project complexity and systems integration: Constructing the London 2012 Olympics and Paralympics Games. International Journal of Project Management.
de Bruijn, K. (2005) Resilience and flood risk management: a systems approach applied to lowland rivers. PhD dissertation. TU Delft. Delft University Press, Delft, Netherlands.
de Bruin, D., Hamhuis, D., van Nieuwenhuijze, L., Overmars, W., Sijmons, D., Vera, F. (1987) Ooievaar: de toekomst van het rivierengebied. Stichting Gelderse Milieufederatie, Arnhem.
de Haan, J. (2006) How emergence arises. Ecological Complexity 3, 293-301.
de Haan, J., Rotmans, J. (2011) Patterns in transitions: Understanding complex chains of change. Technological Forecasting and Social Change 78, 90-102.
DEFRA, (2007) Making space for water update. Department for Environment, Food and Rural Affairs, London, UK.
Degenne, A., Forsé, M. (1999) Introducing social networks. Sage Publications Ltd.
Deltacommissaris, (2011) Deltaprogramma 2012 - Werk aan de delta. Ministerie van Infrastructuur en Milieu, Ministerie van Economische Zaken, Landbouw en Innovatie, The Hague, Netherlands.
Deltacommissie (2008) Samen werken met water - Een land dat leeft, bouwt aan zijn toekomst.
Dietz, T., Ostrom, E., Stern, P. (2003) The struggle to govern the commons. Science 302, 1907.
Downs, P.W., Gregory, K.J., Brookes, A. (1991) How integrated is river basin management? Environmental Management 15, 299-309.
Dryzek, J.S. (1993) Policy analysis and planning: from science to argument. The argumentative turn in policy analysis and planning, 213-232.
Ebbin, S.A. (2002) Enhanced fit through institutional interplay in the Pacific Northwest Salmon co-management regime. Marine Policy 26, 253-259.
Ebregt, J., Eijgenraam, C., Stolwijk, H. (2005) Kosten-baten analyse voor Ruimte voor de Rivier, deel 2 - Kosteneffectiviteit van maatregelen en pakketten. Centraal Plan Bureau, The Hague.

Edelenbos, J., Klijn, E.H. (2006) Managing stakeholder involvement in decision making: A comparative analysis of six interactive processes in the Netherlands. Journal of Public Administration Research and Theory 16, 417-446.

Edelenbos, J., van Buuren, A., van Schie, N. (2011) Co-producing knowledge: joint knowledge production between experts, bureaucrats and stakeholders in Dutch water management projects. Environmental Science & Policy 14, 675-684.

EEA, (2012) Climate change, impacts and vulnerability in Europe 2012. European Environment Agency, Copenhagen.

Eijgenraam, C. (2005) Kosten-batenanalyse voor Ruimte voor de Rivier, deel 1 - Veiligheid tegen overstromen. Centraal Plan Bureau, The Hague.

Ekstrom, J.A., Young, O.R. (2009) Evaluating functional fit between a set of institutions and ecosystems. Ecology and Society 14, 16.

Elzen, B., Wieczorek, A. (2005) Transitions towards sustainability through system innovation. Technological Forecasting and Social Change 72, 651-661.

Ernstson, H., Sörlin, S., Elmqvist, T. (2008) Social movements and ecosystem services-the role of social network structure in protecting and managing urban green areas in Stockholm. Ecology and Society 13, 39.

European Commission, (2013) An EU Strategy on adaptation to climate change, in: Communication from the commission to the european parliament, t.c., the european economic and social committee and the committee of the regions (Ed.), Brussels, Belgium.

Evans, E., Ashley, R., Hall, J., Penning-Rowsell, E., Saul, A., Sayers, P., Thorne, C., Watkinson, A., (2004) Foresight Future flooding. Scientific summary: Volume II Managing future risks. Office of Science and Technology, London, UK.

Farrelly, M., Brown, R. (2011) Rethinking urban water management: Experimentation as a way forward? Global Environmental Change 21, 721-732.

Farrelly, M., Rijke, J., Brown, R., (2012) Exploring operational attributes of governance for change, 7th International Conference on Water Sensitive Urban Design, Melbourne, Australia.

Faust, K., Willert, K., Rowlee, D., Skvoretz, J. (2002) Scaling and statistical models for affiliation networks: patterns of participation among Soviet politicians during the Brezhnev era. Social Networks 24, 231-259.

Feldman, M.S., Orlikowski, W.J. (2011) Theorizing practice and practicing theory. Organization Science 22, 1240-1253.

Ferns, D.C. (1991) Developments in programme management. International Journal of Project Management 9, 148-156.

Feyen, L., Watkiss, P., (2011) The Impacts and Economic Costs of River Floods in Europe, and the Costs and Benefits of Adaptation. Results from the EC RTD ClimateCost Project. , in: Watkiss, P. (Ed.), The ClimateCost Project. Final Report. Stockholm Environment Institute, Stockholm, Sweden.

Flyvbjerg, B., (2007) Truth and Lies About Megaprojects. TU Delft, Delft, Netherlands.

Flyvbjerg, B., Bruzelius, N., Rothengatter, W. (2003) Megaprojects and risk: An anatomy of ambition. Cambridge University Press.

Folke, C. (2003) Freshwater for resilience: a shift in thinking. Philosophical Transactions of the Royal Society of London. Series B: Biological Sciences 358, 2027.

Folke, C. (2006) Resilience: The emergence of a perspective for social–ecological systems analyses. Global Environmental Change 16, 253-267.

Folke, C., Hahn, T., Olsson, P., Norberg, J. (2005) Adaptive governance of social-ecological systems. Annual Review of Environment and Resources 30, 441.

Folke, C., Pritchard, L., Berkes, F., Colding, J., Svedin, U., (1998) The problem of fit between ecosystems and institutions, IHDP Working Paper No. 2, in: Change, I.H.D.P.o.G.E. (Ed.), Bonn, Germany.

Folke, C., Pritchard, L., Berkes, F., Colding, J., Svedin, U. (2007) The problem of fit between ecosystems and institutions: ten years later. Ecology and Society 12, 30.

Foxon, T.J., Reed, M.S., Stringer, L.C. (2009) Governing long term social–ecological change: what can the adaptive management and transition management approaches learn from each other? Environmental Policy and Governance 19, 3-20.

Frank, K., Yasumoto, J. (1998) Linking action to social structure within a system: social capital within and between subgroups. American journal of sociology 104, 642-686.

Galaz, V., Hahn, T., Olsson, P., Folke, C., Svedin, U., (2008) The problem of fit among biophysical systems, environmental and resource regimes, and broader governance systems: Insights and emerging challenges., in: Young, O.R., King, L.A., Schröder, H. (Eds.), Institutions and Environmental Change - Principal Findings, Applications, and Research Frontiers. MIT Press, Cambridge, USA, pp. 147-182.

Geels, F.W. (2002) Technological transitions as evolutionary reconfiguration processes: a multi-level perspective and a case-study. Research Policy 31, 1257-1274.

Geels, F.W., Schot, J. (2007) Typology of sociotechnical transition pathways. Research Policy 36, 399-417.

Gersonius, B., Ashley, R., Pathirana, A., Zevenbergen, C. (2010) Managing the flooding system's resiliency to climate change. Proceedings of the ICE-Engineering Sustainability 163, 15-23.
Gersonius, B., Nasruddin, F., Ashley, R., Jeuken, A., Pathirana, A., Zevenbergen, C. (2012) Developing the evidence base for mainstreaming adaptation of stormwater systems to climate change. Water Research.
Global Water Partnership, (2000) Integrated water resources management, TAC Background Papers. Global Water Partnership Secretariat, Stockholm, Sweden.
Global Water Partnership, (2002) Introducing effective water governance, GWP Technical Paper, Stockholm, Sweden.
Gomez, P.-Y., Jones, B.C. (2000) Crossroads—Conventions: An Interpretation of Deep Structure in Organizations. Organization Science 11, 696-708.
Granovetter, M. (1973) The strength of weak ties. American journal of sociology 78, 1360-1380.
Gray, R.J. (1997) Alternative approaches to programme management. International Journal of Project Management 15, 5-9.
Grindle, M.S. (2004) Good enough governance: poverty reduction and reform in developing countries. Governance 17, 525-548.
Gunderson, L. (1999) Resilience, flexibility and adaptive management--antidotes for spurious certitude? Conservation ecology 3, 1.
Gunderson, L.H., Holling, C.S. (2002) Panarchy: understanding transformations in human and natural systems. Island Pr.
Hahn, T., Olsson, P., Folke, C., Johansson, K. (2006) Trust-building, knowledge generation and organizational innovations: the role of a bridging organization for adaptive comanagement of a wetland landscape around Kristianstad, Sweden. Human Ecology 34, 573-592.
Hajer, M.A. (2005) Setting the stage - A Dramaturgy of Policy Deliberation. Administration & Society 36, 624-647.
Hanf, K., Scharpf, F. (1978) Interorganizational policy making: limits to coordination and central control. Sage Publications.
Harding, R. (2006) Ecologically sustainable development: origins, implementation and challenges. Desalination 187, 229-239.
Heising, W. (2012) The integration of ideation and project portfolio management—A key factor for sustainable success. International Journal of Project Management 30, 582-595.
Herrfahrdt-Pähle, E., Pahl-Wostl, C. (2012) Continuity and change in social-ecological systems: the role of institutional resilience. Ecology and Society 17.
Hertogh, M.J.C.M., Baker, S., Staal-Ong, P.L., Westerveld, E., (2008) Managing large infrastructure projects; research based on Best Practices and

Lessons Learnt In Large Infrastructure Projects in Europe. AT Osborne, Baarn, Netherlands.
Hertogh, M.J.C.M., Westerveld, E., (2010) Playing with Complexity. Management and organisation of large infrastructure projects, PhD thesis. Erasmus University Rotterdam.
Holling, C.S. (1973) Resilience and stability of ecological systems. Annual review of ecology and systematics 4, 1-23.
Holling, C.S., Gunderson, L.H. (2002) Resilience and adaptive cycles. Panarchy: Understanding transformations in human and natural systems, 25-62.
Hood, C. (1991) A public management for all seasons? Public Administration 69, 3-19.
Hooghe, L., Marks, G. (2003) Unraveling the Central State, But How?: Types of Multi-level Governance. American Political Science Review 97, 233-243.
Hooijer, A., Klijn, F., Pedroli, G.B.M., Van Os, A.G. (2004) Towards sustainable flood risk management in the Rhine and Meuse river basins: synopsis of the findings of IRMA-SPONGE. River research and applications 20, 343-357.
Howe, C., Mitchell, C. (2011) Water sensitive cities. IWA Publishing, London, UK.
Hoyer, J., Dickhaut, W., Kronawitter, L., Weber, B. (2011) Water Sensitive Urban Design: Principles and Inspiration for Sustainable Stormwater Management in the City of the Future. Jovis, Hamburg, Germany.
Hu, Y., Chan, A., Le, Y., (2012) Conceptual Framework Of Program Organization For Managing Construction Megaprojects–Chinese Client's Perspective, Engineering Project Organizations Conference, Rheden, Netherlands.
Huisman, P. (2004) Water in the Netherlands: managing checks and balances. Netherlands Hydrological Society (NHV), Utrecht, Netherlands.
Huitema, D., Mostert, E., Egas, W., Moellenkamp, S., Pahl-Wostl, C., Yalcin, R. (2009) Adaptive water governance: assessing the institutional prescriptions of adaptive (co-) management from a governance perspective and defining a research agenda. Ecology and Society 14, 26.
Hulsker, W., Wienhoven, M., van Diest, M., Buijs, S., (2011) Evaluatie ontwerpprocessen Ruimte voor de Rivier. Ecorys.
Huntjens, P., Lebel, L., Pahl-Wostl, C., Camkin, J., Schulze, R., Kranz, N. (2012) Institutional design propositions for the governance of adaptation to climate change in the water sector. Global Environmental Change 22, 67-81.

Imperial, M. (1999) Institutional analysis and ecosystem-based management: the institutional analysis and development framework. Environmental Management 24, 449-465.

Ison, R., Röling, N., Watson, D. (2007) Challenges to science and society in the sustainable management and use of water: investigating the role of social learning. Environmental Science & Policy 10, 499-511.

Jaspers, F.G.W. (2003) Institutional arrangements for integrated river basin management. Water policy 5, 77-90.

Jeffrey, P., Seaton, R. (2004) A conceptual model of 'receptivity'applied to the design and deployment of water policy mechanisms. Journal of Integrative Environmental Sciences 1, 277-300.

Jonas, D. (2010) Empowering project portfolio managers: How management involvement impacts project portfolio management performance. International Journal of Project Management 28, 818-831.

Jones, L., Boyd, E. (2011) Exploring social barriers to adaptation: Insights from Western Nepal. Global Environmental Change 21, 1262-1274.

Juhola, S., Westerhoff, L. (2011) Challenges of adaptation to climate change across multiple scales: a case study of network governance in two European countries. Environmental Science & Policy 14, 239-247.

Kabat, P., Fresco, L.O., Stive, M.J.F., Veerman, C.P., van Alphen, J.S.L.J., Parmet, B.W.A.H., Hazeleger, W., Katsman, C.A. (2009) Dutch coasts in transition. Nature Geoscience 2, 450-452.

Kerzner, H. (2009) Project management: a systems approach to planning, scheduling, and controlling. Wiley.

Kim, K.N., Choi, J.-h. (2013) Breaking the vicious cycle of flood disasters: Goals of project management in post-disaster rebuild projects. International Journal of Project Management 31, 147-160.

Kiser, L., Ostrom, E., (1982) The Three Worlds of Action: A Metatheoretical Synthesis of Institutional Approaches, in: Ostrom, E. (Ed.), Strategies of Political Inquiry. Sage, Beverly Hills, CA, pp. 179-222.

Kjær, A. (2004) Governance, Cambridge: Polity.

Klein, R.J.T., Juhola, S., (2013) A framework for Nordic actor-oriented climate adaptation research, NORD-STAR Working Paper 2013-1. Nordic Centre for Excellence for Strategic Adaptation Research.

Klijn, E.H., Teisman., G.R., (1997) Strategies and games in networks, in: Kickert, W.J.M., Klijn, E.H., Koppenjan, J.F.M. (Eds.), In Managing complex networks: Strategies for the public sector. Sage, London, pp. 98–118.

Klijn, F., de Bruin, D., de Hoog, M.C., Jansen, S., Sijmons, D.F. (2013) Design quality of Room-for-the-River measures in the Netherlands: role and assessment of the Quality Team (Q-team). International Journal of River Basin Management, 1-23.

Knoester, M., Visser, J., Bannink, B., Colijn, C., Broeders, W. (1984) The eastern Scheldt project. Water Science & Technology 16, 51-77.

Kooiman, J. (1993) Modern governance: new government-society interactions. Sage Publications Ltd, London, UK.

Kovats, R.S., Lloyd, S., Hunt, A., Watkiss, P., (2011) The Impacts and Economic Costs on Health in Europe and the Costs and Benefits of Adaptation. Results of the EC RTD ClimateCost Project., in: Watkiss, P. (Ed.), The ClimateCost Project. Final Report. Volume 1: Europe. Stockholm Environment Institute, Stockholm, Sweden.

Kwadijk, J.C.J., Haasnoot, M., Mulder, J.P.M., Hoogvliet, M.M.C., Jeuken, A.B.M., van der Krogt, R.A.A., van Oostrom, N.G.C., Schelfhout, H.A., van Velzen, E.H., van Waveren, H., de Wit, M.J.M. (2010) Using adaptation tipping points to prepare for climate change and sea level rise: a case study in the Netherlands. Wiley Interdisciplinary Reviews: Climate Change 1, 729-740.

Kwak, Y.H., Walewski, J., Sleeper, D., Sadatsafavi, H. (2013) What can we learn from the Hoover Dam project that influenced modern project management? International Journal of Project Management.

Lambin, E. (2005) Conditions for sustainability of huma environment systems: Information, motivation, and capacity. Global Environmental Change 15, 177-180.

Leavitt, H. (1951) Some effects of certain communication patterns on group performance. Journal of abnormal and social psychology 46, 38-50.

Lichtenstein, B., Plowman, D. (2009) The leadership of emergence: A complex systems leadership theory of emergence at successive organizational levels. The Leadership Quarterly 20, 617-630.

Lichtenstein, B., Uhl-Bien, M., Marion, R., Seers, A., Orton, J., Schreiber, C. (2006) Complexity leadership theory: An interactive perspective on leading in complex adaptive systems. Emergence: Complexity & Organization 8, 2.

Lieberherr, E. (2011) Regionalization and water governance: a case study of a Swiss wastewater utility. Procedia-Social and Behavioral Sciences 14, 73-89.

Little, L., McDonald, A. (2007) Simulations of agents in social networks harvesting a resource. Ecological Modelling 204, 379-386.

Loorbach, D. (2010) Transition Management for Sustainable Development: A Prescriptive, Complexity Based Governance Framework. Governance 23, 161-183.

Loorbach, D.A., (2007) Transition management: new mode of governance for sustainable development. Erasmus University, Rotterdam.

Loucks, D.P. (2000) Sustainable water resources management. Water international 25, 3-10.

Lycett, M., Rassau, A., Danson, J. (2004) Programme management: a critical review. International Journal of Project Management 22, 289-299.
Maksimovic, C., Tejada-Guilbert, J. (2001) Frontiers in urban water management: Deadlock or hope. Intl Water Assn.
Marion, R., (2008) Complexity theory for organizations and organizational leadership., in: Uhl-Bien, M., Marion, R. (Eds.), Complexity leadership, part 1: Conceptual foundations. Information Age Publishing Inc., Charlotte, North Carolina, pp. 1-16.
Marion, R., Uhl-Bien, M. (2001) Leadership in complex organizations. The Leadership Quarterly 12, 389-418.
Martinsuo, M., Lehtonen, P. (2007) Role of single-project management in achieving portfolio management efficiency. International Journal of Project Management 25, 56-65.
May, J.V., Wildavsky, A.B., (1978) The Policy Cycle. Sage, Beverly Hill, CA, USA.
Maylor, H., Brady, T., Cooke-Davies, T., Hodgson, D. (2006) From projectification to programmification. International Journal of Project Management 24, 663-674.
McGinnis, M. (1999) Polycentric governance and development: readings from the Workshop in Political Theory and Policy Analysis. Univ of Michigan Pr.
McGinnis, M.D. (2011) An Introduction to IAD and the Language of the Ostrom Workshop: A Simple Guide to a Complex Framework. Policy Studies Journal 39, 169-183.
McLaughlin, M.W. (1987) Learning from Experience: Lessons from Policy Implementation. Educational Evaluation and Policy Analysis 9, 171-178.
Meek, C.L. (2012) Forms of collaboration and social fit in wildlife management: A comparison of policy networks in Alaska. Global Environmental Change.
Milly, P.C.D., Betancourt, J., Falkenmark, M., Hirsch, R.M., Kundzewicz, Z.W., Lettenmaier, D.P., Stouffer, R.J. (2008) Stationarity Is Dead: Whither Water Management? Science 319, 573-574.
Mintzberg, H. (1994) Rise and fall of strategic planning. The Free Press, New York, NY.
Mintzberg, H., Waters, J.A. (1985) Of strategies, deliberate and emergent. Strategic management journal 6, 257-272.
Mitchell, B. (2005) Integrated water resource management, institutional arrangements, and land-use planning. Environment and Planning A 37, 1335-1352.
Mitchell, V. (2006) Applying Integrated Urban Water Management Concepts: A Review of Australian Experience. Environmental Management 37, 589-605.

Morison, P., (2010) Management of Urban Stormwater: Advancing Program Design and Evaluation, School of Geography and Environmental Science. Monash University, Melbourne.
Moser, S.C., Ekstrom, J.A. (2010) A framework to diagnose barriers to climate change adaptation. Proceedings of the National Academy of Sciences 107, 22026-22031.
Moss, T. (2004) The governance of land use in river basins: prospects for overcoming problems of institutional interplay with the EU Water Framework Directive. Land Use Policy 21, 85-94.
Müller, R., Martinsuo, M., Blomquist, T. (2008) Project portfolio control and portfolio management performance in different contexts. Project Management Journal 39, 28-42.
Murray-Webster, R., Thiry, M., (2000) Managing programmes of projects, in: Turner, J.R., Simister, S.J. (Eds.), Gower Handbook of Project Management. Gower Publishing, Aldershot, UK, pp. 33-46.
National Water Commission, (2009) Australian Water Reform 2009: Second biennial assessment of progress in implementation of the National Water Initiative, in: NWC (Ed.), Canberra, ACT, Australia.
National Water Commission, (2011) Urban water in Australia: future directions, in: NWC (Ed.), Canberra, ACT, Australia.
Nelson, D.R., Adger, W.N., Brown, K. (2007) Adaptation to environmental change: contributions of a resilience framework. Annual Review of Environment and Resources 32, 395.
Nelson, R., Howden, M., Smith, M.S. (2008) Using adaptive governance to rethink the way science supports Australian drought policy. Environmental Science & Policy 11, 588-601.
Newig, J., Fritsch, O. (2009) Environmental governance: participatory, multi-level–and effective? Environmental Policy and Governance 19, 197-214.
NSW Office of Water, (2010) 2010 Metropolitan Water Plan, Sydney.
NWC, (2007) National Water Initiative: First Biennial Assessment of Progress in Implementation. National Water Commission, Canberra, ACT.
OECD, (2011) Water Governance in OECD Countries: A Multi-level Approach, OECD Studies on Water. OECD Publishing.
Office for Water Security, (2010) Water for Good: A plan to ensure our water future to 2050. Office for Water Security, Adelaide.
Oh, H., Chung, M., Labianca, G. (2004) Group social capital and group effectiveness: The role of informal socializing ties. The Academy of Management Journal 47, 860-875.
Olsson, P., Folke, C., Berkes, F. (2004a) Adaptive comanagement for building resilience in social–ecological systems. Environmental Management 34, 75-90.

Olsson, P., Folke, C., Galaz, V., Hahn, T., Schultz, L. (2007) Enhancing the fit through adaptive co-management: creating and maintaining bridging functions for matching scales in the Kristianstads Vattenrike Biosphere Reserve Sweden. Ecology and Society 12, 28.
Olsson, P., Folke, C., Hahn, T. (2004b) Social-ecological transformation for ecosystem management: the development of adaptive co-management of a wetland landscape in southern Sweden. Ecology and Society 9, 2.
Olsson, P., Gunderson, L., Carpenter, S., Ryan, P., Lebel, L., Folke, C., Holling, C. (2006) Shooting the rapids: navigating transitions to adaptive governance of social-ecological systems. Ecology and Society 11, 18.
Oostdam, J., Nieuwkamer, R., Brunner, C. (2000) Water: Studenr element in de ruimtelijke ordening? ROM Magazine 20, 30-31.
Opperman, J.J., Galloway, G.E., Fargione, J., Mount, J.F., Richter, B.D., Secchi, S. (2009) Sustainable floodplains through large-scale reconnection to rivers. Science 326, 1487.
Ostrom, E. (1990) Governing the commons: The evolution of institutions for collective action. Cambridge Univ Pr.
Ostrom, E. (1996) Crossing the great divide: coproduction, synergy, and development. World Development 24, 1073-1087.
Ostrom, E. (2007) A diagnostic approach for going beyond panaceas. Proceedings of the National Academy of Sciences 104, 15181.
Ostrom, E. (2011) Background on the Institutional Analysis and Development Framework. Policy Studies Journal 39, 7-27.
Ostrom, E., Cox, M. (2010) Moving beyond panaceas: a multi-tiered diagnostic approach for social-ecological analysis. Environmental Conservation 37, 451-463.
Ostrom, E., Janssen, M.A., Anderies, J.M. (2007) Going beyond panaceas. Proceedings of the National Academy of Sciences 104, 15176.
Paavola, J. (2007) Institutions and environmental governance: A reconceptualization. Ecological Economics 63, 93-103.
Paavola, J., Gouldson, A., Kluvánková-Oravská, T. (2009) Interplay of actors, scales, frameworks and regimes in the governance of biodiversity. Environmental Policy and Governance 19, 148-158.
Padgett, J., Ansell, C. (1993) Robust Action and the Rise of the Medici, 1400-1434. American journal of sociology 98, 1259-1319.
Page, S. (2008) The difference: how the power of diversity creates better groups, firms, schools, and societies. Princeton University Press.
Pahl-Wostl, C. (2006) The importance of social learning in restoring the multifunctionality of rivers and floodplains. Ecology and Society 11, 10.
Pahl-Wostl, C. (2007) Transitions towards adaptive management of water facing climate and global change. Integrated Assessment of Water Resources and Global Change, 49-62.

Pahl-Wostl, C. (2009) A conceptual framework for analysing adaptive capacity and multi-level learning processes in resource governance regimes. Global Environmental Change 19, 354-365.
Pahl-Wostl, C., Craps, M., Dewulf, A., Mostert, E., Tabara, D., Taillieu, T. (2007) Social learning and water resources management. Ecology and Society 12, 5.
Pahl-Wostl, C., Holtz, G., Kastens, B., Knieper, C. (2010) Analyzing complex water governance regimes: the Management and Transition Framework. Environmental Science & Policy 13, 571-581.
Parry, M.L., Canziani, O.F., Palutikof, J.P., van der Linden, P.J., Hanson, C.E. (2007) Climate change 2007: impacts, adaptation and vulnerability. Contribution of Working Group II to the Fourth Assessment Report of the Intergovernmental Panel on Climate Change. IPCC, Cambridge University Press, Cambridge, UK.
Partington, D., (2000) Implementing strategy through programmes of projects, in: Turner, J.R., Simister, S.J. (Eds.), Gower Handbook of Project Management. Gower Publishing, Aldershot, UK, pp. 33-46.
Partington, D., Pellegrinelli, S., Young, M. (2005) Attributes and levels of programme management competence: an interpretive study. International Journal of Project Management 23, 87-95.
PDR, (2011a) 18e Voortgangsrapportage Ruimte voor de Rivier 1 januari 2011 - 30 juni 2011.
PDR, (2011b) 19e Voortgangsrapportage Ruimte voor de Rivier 1 juli 2011 - 30 december 2011.
PDR, (2013) 22ste Voortgangsrapportage Ruimte voor de Rivier 1 januari 2013 - 30 juni 2013.
Pellegrinelli, S. (1997) Programme management: organising project-based change. International Journal of Project Management 15, 141-149.
Pellegrinelli, S. (2002) Shaping context: the role and challenge for programmes. International Journal of Project Management 20, 229-233.
Pellegrinelli, S. (2011) What's in a name: Project or programme? International Journal of Project Management 29, 232-240.
Pellegrinelli, S., Partington, D., Hemingway, C., Mohdzain, Z., Shah, M. (2007) The importance of context in programme management: An empirical review of programme practices. International Journal of Project Management 25, 41-55.
Pellizzoni, L. (2003) Uncertainty and participatory democracy. Environmental Values 12, 195-224.
Pierre, J., Peters, B. (2000) Governance, politics and the state. St. Martin's Press, New York, USA.

Platje, A., Seidel, H. (1993) Breakthrough in multiproject management: how to escape the vicious circle of planning and control. International Journal of Project Management 11, 209-213.
Plowman, D., Solansky, S., Beck, T., Baker, L., Kulkarni, M., Travis, D. (2007) The role of leadership in emergent, self-organization. The Leadership Quarterly 18, 341-356.
Porter, M., Mucha, P., Newman, M., Warmbrand, C. (2005) A network analysis of committees in the US House of Representatives. Proceedings of the National Academy of Sciences of the United States of America 102, 7057.
Project Management Institute (2008a) A Guide to the Project Management Body of Knowledge, Fourth Edition ed. PMI, Newton Square, PA, USA.
Project Management Institute (2008b) The standard for programme management, Second Edition ed. PMI, Newton Square, PA, USA.
Reagans, R., McEvily, B. (2003) Network structure and knowledge transfer: The effects of cohesion and range. Administrative Science Quarterly 48, 240-267.
Reed, M.S. (2008) Stakeholder participation for environmental management: a literature review. Biological conservation 141, 2417-2431.
Reiss, G., Anthony, M., Chapman, J., Leigh, G., Pyne, A., Rayner, P. (2006) Handbook of Programme Management. Gower Publishing Company, Limited, Aldershot, UK.
Rhodes, R. (1996) The New Governance: Governing without Government. Political studies 44, 652-667.
Ribbers, P., Schoo, K.C. (2002) Program management and complexity of ERP implementations. Engineering Management Journal 14, 45-52.
Rijke, J., Brown, R., Zevenbergen, C., Ashley, R., Farrelly, M., van Herk, S., Morison, P. (2012a) Fit-for-purpose governance: A framework to operationalise adaptive governance. Environmental Science & Policy 22, 73-84.
Rijke, J., Farrelly, M., Brown, R., Zevenbergen, C., (2012b) Creating water sensitive cities in Australia: the strengths and weaknesses of current governance approaches, 7th International Conference of Water Sensitive Urban Design, Melbourne, Australia.
Rijke, J., Farrelly, M., Brown, R., Zevenbergen, C. (2013) Configuring transformative governance to enhance resilient urban water systems. Environmental Science and Policy 25, 62-72.
Rijke, J., Farrelly, M., Morison, P., Brown, R., Zevenbergen, C., (2011) Towards improved urban water governance in Adelaide, Australia, 12th International Conference on Urban Drainage, Porto Alegre, Brazil.

Rijke, J., van Herk, S., Zevenbergen, C., Ashley, R. (2012c) Room for the River: Delivering integrated river basin management in the Netherlands. International Journal of River Basin Management 10, 369-382.
Rijke, J., van Herk, S., Zevenbergen, C., Ashley, R., Hertogh, M., Ten Heuvelhof, E. (in press) Adaptive programme management through a balanced performance/strategy oriented focus. International Journal of Project Management.
Rijke, J., Zevenbergen, C., Veerbeek, W., (2009) State of the art Klimaat in de Stad.
Rip, A., Kemp, R. (1998) Technological Change. In: Rayner S., Malone EL (editors).
Ritson, G., Johansen, E., Osborne, A. (2011) Successful programs wanted: Exploring the impact of alignment. Project Management Journal 43, 21-36.
Rockström, J. (2003) Resilience building and water demand management for drought mitigation. Physics and Chemistry of the Earth, Parts A/B/C 28, 869-877.
Rojas, R., Feyen, L., Watkiss, P. (2013) Climate Change and River Floods in the European Union: Socio-Economic Consequences and the Costs and Benefits of Adaptation. Global Environmental Change, in review.
Rotmans, J., Kemp, R., van Asselt, M. (2001) More evolution than revolution: transition management in public policy. Foresight 3, 15-31.
Ruef, M. (2002) Strong ties, weak ties and islands: structural and cultural predictors of organizational innovation. Industrial and Corporate Change 11, 427.
Saeijs, H. (1991) Integrated water management: a new concept. From treating of symptoms towards a controlled ecosystem management in the Dutch Delta. Landscape and Urban Planning 20, 245-255.
Sanderson, J. (2012) Risk, uncertainty and governance in megaprojects: A critical discussion of alternative explanations. International Journal of Project Management 30, 432-443.
Savenije, G. (2009) HESS Opinions'The art of hydrology'. Hydrology and Earth System Sciences 13, 157-161.
Scheffer, M. (2009) Critical transitions in nature and society. Princeton Univ Pr.
Schneider, M., Somers, M. (2006) Organizations as complex adaptive systems: implications of complexity theory for leadership research. The Leadership Quarterly 17, 351-365.
Scott, W.R. (2001) Institutions and organizations. Sage Publications, Inc, Thousand Oaks, USA.

Senge, P., Scharmer, C. (2001) Community action research: learning as a community of practitioners, consultants and researchers. Handbook of Action Research. The Concise Paperback Edition.
Shao, J., Müller, R. (2011) The development of constructs of program context and program success: A qualitative study. International Journal of Project Management 29, 947-959.
Shao, J., Müller, R., Turner, J.R. (2012) Measuring program success. Project Management Journal 43, 37-49.
Shehu, Z., Akintoye, A. (2009) The critical success factors for effective programme management: a pragmatic approach. The Built & Human Environment Review 2, 1-24.
Shkaruba, A., Kireyeu, V. (2012) Recognising ecological and institutional landscapes in adaptive governance of natural resources. Forest Policy and Economics.
Simm, J.D. (2012) A framework for valuing the human dimensions of engineered systems. Proceedings of the ICE-Engineering Sustainability 165, 175-189.
Smith, A., Stirling, A. (2010) The politics of social-ecological resilience and sustainable socio-technical transitions. Ecology and Society 15, 11.
Smith, A., Stirling, A., Berkhout, F. (2005) The governance of sustainable socio-technical transitions. Research Policy 34, 1491-1510.
Taskforce HWBP, (2010) Een dijk van een programma - Naar een nieuwe aanpak van het Hoogwaterbeschermingsprogramma.
Taylor, A., Cocklin, C., Brown, R., Wilson-Evered, E. (2011) An investigation of champion-driven leadership processes. The Leadership Quarterly 22, 412-433.
Teisman, G.R. (2005) Publieke management op de grens van chaos en orde: over leidinggeven en organiseren in complexiteit. Academic Service, Den Haag, Netherlands.
Teller, J., Unger, B.N., Kock, A., Gemünden, H.G. (2012) Formalization of project portfolio management: the moderating role of project portfolio complexity. International Journal of Project Management 30, 596-607.
ten Heuvelhof, E., de Bruijn, H., de Wal, M., Kort, M., van Vliet, M., Noordink, M., Bohm, M., (2007) Procesevaluatie Totstandkoming PKB Ruimte voor de Rivier. Berenschot, Utrecht.
Thiry, M. (2002) Combining value and project management into an effective programme management model. International Journal of Project Management 20, 221-227.
Thiry, M., (2004) Program management: A strategic decision management process, in: Morris, P.W.G., Pinto, J.K. (Eds.), The Wiley guide to managing projects. John Wiley & Sons Inc, Hoboken, NJ, USA, pp. 257-287.

Tompkins, E.L., Adger, W.N. (2004) Does adaptive management of natural resources enhance resilience to climate change? Ecology and Society 9, 10.

Too, E.G., Weaver, P. (2013) The management of project management: a conceptual framework for project governance. International Journal of Project Management.

Turner, J.R. (1999) The handbook of project-based management. McGraw-Hill London.

Turner, J.R., (2000) An encyclopedia for the profession of project management, in: Turner, J.R., Simister, S.J. (Eds.), Gower Handbook of Project Management. Gower Publishing, Aldershot, UK, pp. 1-25.

Turnhout, E., Van Bommel, S., Aarts, N. (2010) How participation creates citizens: participatory governance as performative practice. Ecology and Society 15, 26.

Turrini, A., Cristofoli, D., Frosini, F., Nasi, G. (2010) Networking literature about determinants of network effectiveness. Public Administration 88, 528-550.

Uhl-Bien, M., Marion, R. (2009) Complexity leadership in bureaucratic forms of organizing: A meso model. The Leadership Quarterly 20, 631-650.

Uhl-Bien, M., Marion, R., McKelvey, B. (2007) Complexity leadership theory: Shifting leadership from the industrial age to the knowledge era. The Leadership Quarterly 18, 298-318.

UK Office of Government Commerce (2007) Managing successful programmes, Third Edition ed. Stationery Office, London, UK.

Unger, B.N., Gemünden, H.G., Aubry, M. (2012) The three roles of a project portfolio management office: Their impact on portfolio management execution and success. International Journal of Project Management 30, 608-620.

UNISDR (2013) From Shared Risk to Shared Value –The Business Case for Disaster Risk Reduction. Global Assessment Report on Disaster Risk Reduction. United Nations Office for Disaster Risk Reduction (UNISDR), Geneva, Switzerland.

Valk, D., Wolsink, M. (2001) Water als ordende factor. Rooilijn 34, 285-290.

van Aken, T.G.C. (1996) De weg naar projectsucces. Reed Business, Amsterdam, Netherlands.

van Buuren, A., Buijs, J.M., Teisman, G. (2010) Program management and the creative art of coopetition: Dealing with potential tensions and synergies between spatial development projects. International Journal of Project Management 28, 672-682.

van de Meene, S., Brown, R., Farrelly, M. (2011) Towards understanding governance for sustainable urban water management. Global Environmental Change 21, 1117–1127.

van den Brink, M. (2009) Rijkswaterstaat on the horns of a dilemma. Eburon Uitgeverij BV, Delft, Netherlands.
van der Brugge, R., Rotmans, J. (2007) Towards transition management of European water resources. Water Resources Management 21, 249-267.
van der Brugge, R., Rotmans, J., Loorbach, D. (2005) The transition in Dutch water management. Regional Environmental Change 5, 164-176.
van der Brugge, R., van Raak, R. (2007) Facing the adaptive management challenge: insights from transition management. Ecology and Society 12, 33.
van Heezik, A. (2007) Strijd om de rivieren: 200 jaar rivierenbeleid in Nederland. HNT Historische Produkties, Den Haag/Haarlem, Netherlands.
van Herk, S., Herder, P., De Jong, W., Alma, D. (2006) Innovative contracting in infrastructure design and maintenance. International journal of critical infrastructures 2, 187-200.
van Herk, S., Rijke, J., Zevenbergen, C., Ashley, R., (2012a) Governance of integrated flood risk management to deliver large scale investment programmes: delivery-focussed social learning in the Netherlands, FLOODrisk 2012, Rotterdam, Netherlands.
van Herk, S., Rijke, J., Zevenbergen, C., Ashley, R., (2012b) Governance of integrated flood risk management to deliver large scale investment programmes: delivery focused social learning in the Netherlands., Floodrisk 2012 - 2nd European Conference on flood risk management, Rotterdam, Netherlands.
van Herk, S., Rijke, J., Zevenbergen, C., Ashley, R., (2012c) Transition in governance of River Basin Management in the Netherlands through multi-level social learning, International Sustainability Transitions Conference, Copenhagen, Denmark.
van Herk, S., Rijke, J., Zevenbergen, C., Ashley, R., (2012d) Transition in governance of River Basin Management in the Netherlands through social learning, International Sustainability Transitions Conference, Copenhagen, Denmark.
van Herk, S., Rijke, J., Zevenbergen, C., Ashley, R. (2013) Understanding the transition to integrated flood risk management in the Netherlands. Environmental Innovation and Societal Transitions.
van Herk, S., Rijke, J., Zevenbergen, C., Ashley, R. (under review) Attributes for integrated Flood Risk Management projects; case study Room for the River. International Journal of River Basin Management.
van Herk, S., Rijke, J., Zevenbergen, C., Ashley, R., Besseling, B. (in press) Adaptive co-management and network learning in the Room for the River programme. Journal of Environmental Planning and Management.

van Herk, S., Zevenbergen, C., Ashley, R., Rijke, J. (2011a) Learning and Action Alliances for the integration of flood risk management into urban planning: a new framework from empirical evidence from The Netherlands. Environmental Science & Policy 14, 543-554.
van Herk, S., Zevenbergen, C., Rijke, J., Ashley, R. (2011b) Collaborative research to support transition towards integrating flood risk management in urban development. Journal of Flood Risk Management 4, 306-317.
van Nieuwaal, K., Driesen, P., Spit, T., Termeer, K., (2009) A state of the art of governance literature on adaptation to climate change: towards a research agenda
van Stokkom, H.T.C., Smits, A.J.M., Leuven, R.S.E.W. (2005) Flood Defense in The Netherlands. Water international 30, 76-87.
van Twist, M., Ten Heuvelhof, E., Kort, M., Olde Wolbers, M., van den Berg, C., Bressers, N., (2011a) Bijlgenboek Tussenevaluatie Ruimte voor de Rivier.
van Twist, M., Ten Heuvelhof, E., Kort, M., Olde Wolbers, M., van den Berg, C., Bressers, N., (2011b) Tussenevaluatie PKB Ruimte voor de Rivier.
Veerbeek, W., Ashley, R., Zevenbergen, C., Gersonius, B., Rijke, J., (2012) Building adaptive capacity for flood proofing in urban areas through synergistic interventions. , 7th International Conference on Water Sensitive Urban Design, Melbourne, Australia.
Voß, J.P. (2007) Innovation processes in governance: the development of'emissions trading'as a new policy instrument. Science and Public Policy 34, 329-343.
Voß, J.P., Bauknecht, D., Kemp, R. (2006) Reflexive governance for sustainable development. Edward Elgar Publishing.
Voß, J.P., Bornemann, B. (2011) The Politics of Reflexive Governance: Challenges for Designing Adaptive Management and Transition Management. Ecology and Society 16, 9.
Vrijling, H., (2008) Verloren in een zee van mooie plannen, Trouw.
Walker, B., Holling, C.S., Carpenter, S.R., Kinzig, A. (2004) Resilience, Adaptability and Transformability in Social--ecological Systems. Ecology and Society 9, 5.
Warner, J., van Buuren, A., Edelenbos, J., (2013) Making space for the river. IWA Publishing, London, UK.
Wasserman, S., Faust, K. (1994) Social network analysis: Methods and applications. Cambridge Univ Pr.
Webler, T., Tuler, S. (2006) Four perspectives on public participation process in environmental assessment and decision making: Combined results from 10 case studies. Policy Studies Journal 34, 699-722.

Webler, T., Tuler, S., Krueger, R. (2001) What is a good public participation process? Five perspectives from the public. Environmental Management 27, 435-450.
Wesselink, A., Paavola, J., Fritsch, O., Renn, O. (2011) Rationales for public participation in environmental policy and governance: practitioners' perspectives. Environment and Planning-Part A 43, 2688.
Wiering, M.A., Arts, B.J.M., (2006) Discursive shifts in Dutch river management: 'deep' institutional change or adaptation strategy?, in: Leuven, R.S.E.W., Ragas, A.M.J., Smits, A.J.M., Velde, G. (Eds.), Living Rivers: Trends and Challenges in Science and Management. Springer Netherlands, pp. 327-338.
Winsemius, P. (1986) Gast in eigen huis: beschouwingen over milieumanagement. Samsom HD Tjeenk Willink.
WNF, (1992) Levende rivieren, Report Wereld Natuur Fonds.
Wolf, C., Floyd, S.W. (2013) Strategic Planning Research Toward a Theory-Driven Agenda. Journal of Management.
Wolsink, M. (2006) River basin approach and integrated water management: Governance pitfalls for the Dutch Space-Water-Adjustment Management Principle. Geoforum 37, 473-487.
Wong, P., Rann, M., Maywald, K., (2009) Media release - Go ahead for $150 million in stormwater projects, in: Government of South Australia (Ed.), Adelaide, Australia.
Wong, T., Brown, R. (2009) The water sensitive city: principles for practice. Water Science & Technology 60, 673-682.
World Bank (2013) Strong, Safe, and Resilient - A Strategic Policy Guide for Disaster Risk Management in East Asia and the Pacific. World Bank, Washington, DC.
Wright, S.A.L., Fritsch, O. (2011) Operationalising active involvement in the EU Water Framework Directive: Why, when and how? Ecological Economics 70, 2268-2274.
Young, O., Underdal, A., (1997) Institutional dimensions of global change, IHDP Scoping Report, in: International Human Dimensions Programme on Global Environmental Change (Ed.), Bonn, Germany.
Young, R., Young, M., Jordan, E., O'Connor, P. (2012) Is strategy being implemented through projects? Contrary evidence from a leader in New Public Management. International Journal of Project Management 30, 887-900.
Zevenbergen, C., van Herk, S., Rijke, J., Kabat, P., Bloemen, P., Ashley, R., Speers, A., Gersonius, B., Veerbeek, W. (2012) Taming global flood disasters. Lessons learned from Dutch experience. Natural Hazards, 1-9.

Zevenbergen, C., Veerbeek, W., Gersonius, B., Van Herk, S. (2008) Challenges in urban flood management: travelling across spatial and temporal scales. Journal of Flood Risk Management 1, 81-88.

APPENDIX A: FARRELY ET AL (2012)

Farrelly, M., Rijke, J. and Brown, R. (2012) Exploring operational attributes of governance for change, 7th International Conference on Water Sensitive Urban Design, 21-23 February 2012, Melbourne, Australia.

Appendix A: Farrelly et al (2012)

Exploring operational attributes of governance for change

M.A. Farrelly*, J. Rijke*,** and R.R. Brown*

*Centre for Water Sensitive Cities, School of Geography & Environmental Science, Monash University, Wellington Rd, Clayton, Victoria, 3810,Australia (E-mail: megan.farrelly@monash.edu; rebekah.brown@monash.edu)

** Flood Resilience Group, UNESCO-IHE, PO Box 3015, 2601 DA, Delft, Netherlands (E-mail: j.rijke@unesco-ihe.org)

ABSTRACT

Building a pathway to achieve widespread institutionalisation of water sensitive urban design requires, among other elements, a shift in conventional governance and management practices. Current academic scholarship regarding broad governance trends focuses predominantly on appropriate modes and principles for practice, with little attention directed towards developing operational pathways for change. This paper draws on critical insights from three major research phases of a national, social research program undertaken between 2004 and 2011, which collectively aimed to assist urban water managers in generating socio-institutional change towards water sensitive cities. Through detailed historical and contemporary research, which investigated the institutionalisation of water sensitive urban design in Melbourne, the importance of experimentation and learning, and current governance practices in Australian cities, eight key factors for enabling transitions governance were identified. The key factors included: establishing a narrative and metaphor; organisational and individual leadership; policy framework and institutional design; regulatory and compliance agendas; an economic and business case; capacity building and demonstration; public engagement and behaviour change; and, research and policy/practice partnerships. Importantly, these eight areas of operational governance demonstrated a strong interplay between core governance structures and processes; suggesting there is a need to have all factors aligned before a system-wide transition can be successfully expedited. These key operational governance factors provide the basis of a future reform agenda related to mainstreaming water sensitive urban design in Australian cities.

KEYWORDS
governance; socio-institutional change; policy design.

INTRODUCTION

Cities worldwide are facing a variety of internal and external pressures (e.g. increasing populations, climate change) on essential service infrastructure, including water supply, wastewater and drainage. This has led to industry commentators from the social and technical sciences to suggest the conventional linear, large-scale systems (Newman, 2001) are no longer the only solution to providing urban populations with the required potable water, public health protection and flood management services (e.g. Wong, 2006, Brown *et al.*, 2009a). Indeed, numerous technological advances have been made at various scales which can either replace the traditional configuration of urban water management or operate concurrently (e.g. Brown *et al.*, 2009a). Similarly, research suggests that adopting alternative practices in the urban water sector is not constrained by technologies, but rather a suite of socio-institutional barriers which limit the scope of change (Brown *et al.*, 2009b) and contribute towards (technological) path-dependent trajectories (Brown *et al.*, 2011). Thus, the challenge of transforming the urban water sector towards more sustainable practices lies within the realm of governance, as opposed to purely technical change.

Indeed, a number of scholars have commented that the most pressing issue facing resource-based problems is primarily a governance issue (e.g. Adger, 2005; Pahl-Wostl, 2009, OECD, 2011). Governance is not a new concept; over the last two decades, academic scholarship has focused on exploring principles for good governance and on the different modes of delivery such as hierarchies, markets, networks (Elzen and Wieczorek, 2005) and hybrid approaches (van de Meene *et al.*, 2011). Despite this body of knowledge, there has been limited empirical investigation into understanding the critical operational (prescriptive) components of governance practices to help stimulate change in resource-based sectors.

To contribute towards the academic debate, this paper proposes eight operational governance attributes which collectively, are the result of synthesising critical insights from three recent, major research projects embedded within a national research program aimed at improving urban water governance in Australia. The Urban Water Governance Program, situated at Monash University (Melbourne), is actively involved in empirically exploring and testing questions of governance related to promoting change in the Australian urban water sector. This normative research program accepts that change in governance is required to support a shift in urban water technology adoption, management and practices. The paper starts with an overview of recent key governance scholarship insights and outlines the three key research projects and the analytical approach used to identify the operational factors. Following this, the eight core attributes will be presented before a brief discussion regarding the potential role these factors may play in future urban water reform agendas.

GOVERNANCE FOR CHANGE

Governance is not a new concept, but has become more prominent over the last two decades in response to an increase in the number of stakeholders involved in essential service management and delivery. Distinct from government, the term governance captures the dynamic interaction between the processes (e.g. managing networks, markets, communities) and structures (e.g. institutional design of patterns and mechanisms) required to steer and manage society (see e.g. van de Meene *et al.* (2011). Thus, understanding urban water governance requires an exploration of the structures and process which can be purposively enacted to guide, steer, control or manage urban water systems. Recent research has been exploring the concept of adaptive governance, whereby systems are managed through i) anticipating long-term change (i.e. global climate change; population growth), ii) preparedness for responding to system shocks (i.e. drought, flood) and iii) building the capacity to recover from shock events (see e.g. Folke *et al.*, 2005). This approach is increasingly advocated amongst scholars for the approach embraces the inherent complexity and uncertainty in managing resource systems and recognises the on-going interaction across spatial and temporal scales (e.g. Adger *et al.*, 2005). Table 1 provides a synthesis of key principles for adaptive governance identified in the academic scholarship. Although an adaptive governance approach might be 'ideal', in reality it is exceptionally challenging to put into operation (Rijke *et al.*, submitted-a). This may be attributed to: i) the complications arising from contextual nuances; ii) an emphasis on descriptive and analytical theories rather than prescriptive approaches (Loorbach, 2010); iii) the term governance often invokes ambiguity and uncertainty with regard to purpose; iv) unclear governance (operational) contexts; v) uncertain governance outcomes (see Rijke *et al.*, submitted-a). There are, however, recent exceptions including Loorbach (2010), van de Meene *et al.* (2011), Huntjens *et al.* (2011) and Rijke *et al.* (submitted-a).

To improve the uptake of adaptive governance strategies, Rijke *et al.* (submitted-a) have proposed a 'fit-for-purpose' governance framework to assess the level of appropriateness of current and planned governance strategies for handling a particular issue (i.e. protecting waterway health; securing water supplies). The authors argue that using their three-step framework is a practical starting point in helping to establish effective adaptive governance. In a similar premise, recent work by Huntjes *et al.* (2011) have proposed eight broad institutional design propositions in support of adaptive governance.

Table A.0.1 Characteristics of Adaptive Governance

Component	Characteristic	Description
What?	**Anticipation**	Forecasting long term change (i.e. climate change, population growth) and preparedness to immediate shocks (i.e. droughts and floods).
	Reflexivity	Reviewing policies and effectiveness of governance strategies
	Flexibility	The ability to adjust strategies to changing drivers and problems.
	Robustness	Incorporating a degree of redundancy to change by avoiding emotive decision-making and ensuring rational decisions continue to regulate behaviour and provide a predictable arena for interaction.
Who?	**Self-organisation**	Every-day interactions (and those brought about by crises) between individuals and organisations can

		lead to emergent outcomes such as learning, adoption and rejection of new approaches.
	Leadership	Leaders can generate a breakthrough in dominant mindsets, introduce and/or impose visions for the future and strategically bring together people, resources and knowledge.
How?	**Collaborative decision making**	Urban water governance involves many stakeholders. Processes of co-initiation, co-design and co-implementation is encouraged to synthesise and implement different knowledge and experience to avoid inappropriate decisions.
	Multi-level governance	Create nested institutional arrangements amongst national, state, regional and local governments and the private sector, and facilitate formal and informal networks of individuals within and between organisations.
	Research and Development	Testing and applying new solutions is critical for learning processes. Research enhances discovery and understanding, supports capacity building, and helps deliver practical outcomes.
	Data management	Monitoring and evaluation, collection, storage and preparation of data for a range of applications by different users.
Sources: Due to conference page limits the existing scholarship which informs this table can be sourced from the lead author.		

RESEARCH APPROACH

To build our understanding of relevant governance approaches to support a sector-wide transition, this paper synthesises the outcomes from three unique, independent, but highly inter-related research activities, undertaken between 2004 and 2011. These three research projects are part of the Urban Water Governance Program which investigates pathways for institutionalising sustainable urban water management in Australia. Each qualitative, empirical, social science research project is briefly described below. While each project addressed unique research questions, there were similarities across the research design, case study selection, applied methods (Table 2) and key research outcomes, which led the authors to reflect upon how this body of knowledge may contribute towards an understanding of operational governance components required for more sustainable urban water management. Each of the authors has had involvement in at least one of the research projects, while one author was involved in all three projects.

A) Transition to water sensitive urban design – the story of Melbourne (2004-2007)

Drawing on transition theory as an analytical framework, Brown and Clarke (2007) focused on a retrospective analysis of the institutionalisation of urban stormwater quality management in Melbourne. Following interviews (n=52) with leading, senior urban water practitioners and urban stormwater managers, the research explored important factors involved in supporting institutional change, which led towards the mainstreaming of water sensitive urban design approaches in Melbourne, Australia (see Brown and Clarke (2007) for further details).

B) Rethinking urban water management: experimentation as a way forward (2008-2011)

Reflecting on the key institutional change factors identified by Brown and Clarke (2007), Farrelly and Brown (2011) embarked upon an investigation regarding how local-scale experiments (i.e. adoption of alternative technologies and/or practices) were established and maintained despite operating in a predominantly unsympathetic regime. Drawing on transitions management and social learning scholarship the authors undertook an embedded, multiple case study approach (following Yin, 2009), whereby eleven (predominantly technical) local-scale experiments were critically examined. The experiences of key urban water practitioners, alongside the experiment investigations, were collated and closely critiqued, revealing a suite of enabling and constraining factors regarding how these local scale experiments can influence change at both a local and regime scale in current urban water management practices.

C) Examining fit-for-purpose governance in Australia (2010-2011)

Drawing on adaptive governance and transition management scholarship, this project developed a diagnostic framework to assess whether existing governance strategies and responses to exogenous pressures are meeting their intended purpose(s) in different states of Australia. Based on a comparative case study approach (see Yin, 2009) that investigated practitioner insights (n=90) across three cities (see Table 2), this study revealed the strengths and weaknesses of different governance approaches (i.e. centralised vs. decentralised and formal institutions vs. informal networks). Rijke *et al.* (submitted-b) outlines more details regarding research approach and outcomes.

During project C, the authors collectively began to reflect upon the broader cumulative understanding regarding critical operational practices that were emerging from the Urban Water Governance Program's body of research. With this in mind, the authors undertook to re-examine the three existing three data sets regarding institutional regime activity to identify key operational factors (leverage points) which promoted, supported and/or generated change in governance practices. Points of commonality and divergence were sought to test the strength of the explanation surrounding each of the eight identified factors, and contemporary academic scholarship regarding, for example, common pool resource management issues, have been drawn upon where appropriate to elaborate each of the eight factors.

Table A.0.2 Methods used in each research project

Design and Methods	Research Project		
	(A) WSUD in Melbourne	(B) Exploring experimentation	(C) Fit-for-purpose governance
Case study location	Melbourne	Melbourne, Brisbane and Perth	Melbourne, Sydney and Adelaide
Semi-structured interviews	✓ n = 52	✓ n= 154	✓ n=90
Secondary data collection [a]	✓	✓	✓
Validation of key insights	✓	✓	✓

a= sources include policy briefings and statements, newspaper articles, media campaigns, academic literature, professional association and industry reports.

ENABLING GOVERNANCE FOR CHANGE

To address the lack of prescription in governance scholarship, the re-examination of three recent empirical data sets has revealed eight socio-institutional factors considered to have the capacity to influence existing and future governance approaches, and hence the ability of the urban water sector to adopt new practices (i.e. stormwater harvesting, treatment and reuse). These eight factors are outlined briefly in Table 3. All eight factors were identified in all three research projects, with one exception, project B did not identify the importance of public engagement/behaviour change given the focus was on exploring predominantly technical experimentation design to support learning amongst practitioners in the urban water sector. By identifying these operational factors, this research has extended the characteristics of adaptive governance (Table 2) and recent institutional design propositions (see Huntjens *et al.* (2011) to provide more specific direction for decision-makers to consider when designing broad governance reforms.

It is important to recognise that there is significant interplay amongst the factors, particularly when considering the roles of structures and processes. Drawing on Giddens' (1984) theory of structuration, structures are considered to be "rules and resources, recursively implicated in the reproduction of social systems". Hence, the core structural factors identified in this research are relatively stable over long timeframes, but remain subject to reinterpretation through the process factors, which can adapt more readily to changing circumstances (i.e. external and internal events) over shorter timeframes. While broad contextual conditions will exert influence over the trajectory of a transition (i.e. politics, economy, technology), to guide and steer the transition requires explicit processes including leadership (distributed, organisational, individual), dedicated, sectoral capacity building programs, extensive engagement with the broader socio-cultural context and research partnerships (between science, policy and practice) (i.e. technical and governance experimentation) (Table 3). These processes will ultimately influence, but will also be guided by, the core structural attributes identified: vision, narrative, metaphor; policy design and frameworks; economic incentives; and, regulatory and compliance agendas.

Table A.0.3 **Operational factors supporting transition governance**

OPERATIONAL FACTORS	SUB-COMPONENTS
	STRUCTURE
Narrative, metaphor and image (e.g. a clear vision)	– Storyline that invokes a need for change – Visual connection to problems and potential solutions(e.g. Baumgartner and Jones, 1991; Boal and Schultz, 2007; Dryzek, 1993)(e.g. Baumgartner and Jones, 1991; Boal and Schultz, 2007; Dryzek, 1993)(e.g. Baumgartner and Jones, 1991; Boal and Schultz, 2007; Dryzek, 1993)(e.g. Baumgartner and Jones, 1991; Boal and Schultz, 2007; Dryzek, 1993)(e.g. Baumgartner and Jones, 1991; Boal and Schultz, 2007; Dryzek, 1993)(e.g. Baumgartner and Jones, 1991; Boal and Schultz, 2007; Dryzek, 1993)(e.g. Baumgartner and Jones, 1991; Boal and Schultz, 2007; Dryzek, 1993)(e.g. Baumgartner and Jones, 1991; Boal and Schultz, 2007; Dryzek, 1993)
Regulatory and compliance agenda	– Objectives and mechanisms (markets, legislative rules and education) – Performance targets – Monitoring, enforcement and evaluation
Economic justification	– Demonstrated business case – Appropriate allocation/evaluation of all social and environmental costs and benefits (monetary and non-monetary)
Policy & planning frameworks & institutional design	– Define the scope of the policy – Highlight the distribution and trade-offs of costs and benefits – Legislation, administrative organisational arrangements – Dedicated funding streams
	PROCESS
Leadership	– Distributed network leadership (policy, operational, private sector, science, community and political) – Organisational leadership – Positional and personal leadership characteristics
Capacity building and demonstration	– Creating awareness about problems and solutions – Build confidence in approach, technology and practice – Develop new skills and competencies across the sector – Creating informal incentives to apply and replicate leanings
Public engagement and behaviour change	– Understanding existing community drivers – Informing and engaging with the community – Encouraging behaviour change amongst community members
Research and partnerships with policy/practice	– Science partnerships: co-constructing science, policy and practice agendas for evidence-based decision-making

Structures

Loorbach (2010 p.162) asserts that new modes of governance are required to help provide direction and coordination for change. Having a clearly articulated vision is well recognised in the scholarship as being critical to setting a directional path for change (e.g. Wong and Brown, 2009). Having a clear storyline or narrative was considered important in building a common understanding of the project and/or scale of change required in all three research projects. The emphasis is on generating a narrative, metaphor and image (visual connection to problem and potential solutions) to help engage with a broad range of stakeholders and to help them connect to the (often motherhood-type) overall vision. Conceiving of the narrative as a structural element underscores its importance in providing a stable (but evolving) direction for change in urban water management practices.

Given the dominant market-based modes of governance delivery currently operating in Australian urban water management (van de Meene *et al.* 2011), it is essential that a sound economic justification for change in traditional operation can be made. This justification must look beyond the conventional approach of focusing solely on priced impacts to incorporating a value-based evaluation which capture the non-priced benefits and costs of new technologies, practices and approaches. However, this data is difficult to separate out given the high level of interconnectedness between the multiple benefits (i.e. the micro-climatic and aesthetic benefits of well-designed, vegetated, biofiltration systems).

While traditionally perceived as a barrier to change (e.g. Brown *et al.*, 2009b), the regulatory environment, if used in a reflexive manner, can provide a basic level of expected practice within the sector. By setting critical performance targets and objectives, the regulatory framework can also provide direction to relevant stakeholders in the sector (i.e. expectation of incorporating alternative water systems in new and retrofit developments, where appropriate). Therefore, regular reviews are required to ascertain the efficacy of existing framework(s). Similarly, having clearly defined policies and appropriate administrative and organisational arrangements (the 'hard' structures) is important for a coherent and connected operational environment for urban water management. For example, future emphasis should also be directed towards further integration between policy and planning frameworks to ensure the most effective land use planning outcomes. Policy designers, alongside treasury officials, must ensure the perennial challenge of insufficient funds is overcome by setting aside dedicated, ongoing funding streams to ensure funds are available for existing and emergent initiatives. Finally, any changes in either the regulatory environment or policy platforms must be complemented with associated capacity building modules to ensure the relevant skills and competencies are available within the sector to ensure the required changes/adaptations become operational.

Processes

Exerting an influence on the structures, the processes outlined in Table 3 are fundamental in building momentum amongst stakeholders in terms of understanding, knowledge and confidence in the need for and ability to change. In all three studies, dedicated (e.g. secured, on-going funds) capacity building programs were considered vital to promoting confidence in new and emerging technologies and practices, and for providing the necessary skills training required for broad scale adoption (see, e.g. Farrelly and Brown, 2011). Importantly, such programs require a close alignment with current physical, technical and social science research, so that outcomes are readily transformed to influence and facilitate practical change.

Separate to, but closely aligned with the capacity building and demonstration factor, is the requirement for explicit research-policy partnerships. Research in all facets (technical, physical and social) is an ongoing imperative to inform evidence-based policy formulation. However, there remains widespread evidence of policy-practice (implementation) disconnection (see Farrelly and Brown, 2011); thus, the authors contend that through more direct, purposive interaction among policy makers and researchers there will be greater capacity to collectively explore implementation challenges and solutions to inform practical change. Furthermore, process driven change can be facilitated through leadership (see e.g. Taylor, 2009). In particular, project C revealed the importance of distributed leadership (amongst representatives within policy, operations, the private sector, science, politics and the community), as well as organisational leadership (i.e. at the board or executive level) and individual leaders (based on positional and personal characteristics).

CONCLUSIONS

Water crises have been referred to as a crisis of governance, yet there remains limited prescriptive information regarding how to achieve the required change in governance to facilitate the adoption of more sustainable urban water practices. Reflecting upon three unique, but highly interrelated research projects within the Urban Water Governance Program, this paper has identified eight operational attributes supporting transitions governance. These attributes provide some direction to key policy makers in designing future governance reforms related to mainstreaming water sensitive cities and offer a platform for reflection upon current practice to help identify areas in need of vital change and investment. Transforming towards more sustainable urban water infrastructure and servicing will require a number of shifts in governance practice (i.e. from centralised/formal to decentralised/informal or vice versa) depending on the emerging context(s) (social, technical, political, economic). To help generate transformative change, it is important for policy designers to reflect upon existing structural elements to ensure that these are fundamentally capable of supporting a change and then attention can be directed towards promoting broader socio-cultural shifts.

ACKNOWLEDGEMENT

This research was funded by a consortium of public and private organisations who support the research activities of the Cities as Water Supply Catchments research programme situated within the Centre for Water Sensitive Cities, Monash University.

REFERENCES

Adger, W., Huges, T.P., Folke, C., Carpenter, S.R. and Rockstrom, J. (2005). Social-ecological resilience to coastal disasters. Science, 309(5737), 1036.

Brown, R.R. and Clarke, J. (2007). Transition to Water Sensitive Urban Design: The Story of Melbourne, Australia. Report 07/1, Facility for Advancing Water Biofiltration, Monash University, Melbourne.

Brown, R., Keath, N. and Wong, T. (2009a). Urban water management in cities: historical, current and future regimes. Water Science and Technology 59, 847-855.

Brown, R., Farrelly, M. and Keath, N. (2009b). Practitioner perceptions of social and institutional barriers to advancing a diverse water source approach in Australia. International Journal of Water Resources Development, 25(1), 15-28.

Brown, R.R., Ashley, R. and Farrelly, M.A. (2011). Political and professional agency entrapment: an agenda for urban water research. Water Resources Management, 25(15), 4037-4050.

Elzen, B. and Wieczorek, A. (2005). Transitions towards sustainability through system innovation. Technological Forecasting and Social Change, 72, 651-661.

Farrelly, M. and Brown, R. (2011). Rethinking urban water management: Experimentation as a way forward? Global Environmental Change 21, 721-732.

Folke, C., Hahn, T., Olsson, P. and Norberg, J. (2005). Adaptive governance of social-ecological systems. Annual Review of Environment and Resources, 30, 441-473.

Giddens, A. (1984). The Constitution of Society: Outline of the Theory of Structuration. Univeristy of California Press, Berkely, Los Angeles.

Huntjens, P., Lebel, L., Pahl-Wostl, C., Rshulze, R., Camkin, J. and Kranz, N. (2011). Institutional design propositions for the governance of adaptation to climate change in the water sector. Global Environmental Change, doi:10.1016/j.lobenvcha.2011.09.015.

Loorbach, D. (2010). Transition management for sustainable development: a prescriptive, complexity-based governance framework. Governance, 23,161-183.

Newman, P. (2001). Sustainable urban water systems in rich and poor cities – steps towards a new approach. Water Science and Technology, 43, 93-99.

Pahl-Wostl, C. (2009). A conceptual framework for analysing adaptive capacity and multi-level learning processes in
resource governance regimes. Global Environmental Change, 19, 54-365.

Rijke, J., Brown, R., Zevenbergen, C., Ashley, R., Farrelly, M., Morison, P. and van Herk, S. (submitted -a). Fit-for-purpose governance: a framework to make adaptive governance operational. Environmental Science and Policy.

Rijke, J., Farrelly, M., Zevenbergen, C. and Brown, R. (submitted -b) Creating water sensitive cities in Australia: the fit-for-purpose of current governance approaches, 7th International Conference on Water Sensitive Urban Design, Melbourne, Australia.

Taylor, A. (2009). Sustainable urban water management: understanding and fostering champions of change. *Water Science and Technology*, 59(5), 883-891

van de Meene, S.J, Brown, R.R. and Farrelly, M.A. (2011). Towards understanding governance for sustainable urban water management: a practice-oriented perspective. Global Environmental Change, 21 (2011) 1117–1127.

Wong, T.H.F. (2006). Chapter 1 – Introduction. In Australian Runoff Quality: A guide to Water Sensitive Urban Design (THF Wong, ed) Engineers Australia, Canberra.

Yin, R.K. (2009). Case Study Research : Design and Methods, Sage Publications, Thousand Oaks, California.

APPENDIX B: RIJKE ET AL (2012b)

Rijke, J., Farrely, M., Brown, R., Zevenbergen, C. (2012) Creating water sensitive cities in Australia: the strengths and weaknesses of current governance approaches. 7th International Conference on Water Sensitive Urban Design, 21-23 February 2012, Melbourne, Australia.

Appendix B: Rijke et al (2012b)

Creating water sensitive cities in Australia: the strengths and weaknesses of current governance approaches

J. Rijke*, **, M. Farrelly*, R. Brown*, C. Zevenbergen**

* Centre for Water Sensitive Cities, Monash University,VIC 3800, Australia (E-mail: j.rijke@unesco-ihe.org, megan.farrelly@monash.edu, rebekah.brown@monash.edu)
** Flood Resilience Group, UNESCO-IHE, PO Box 3015, 2601 DA, Delft, Netherlands (E-mail: c.zevenbergen@unesco-ihe.org)

ABSTRACT
This paper details the outcomes of a qualitative, social science research project, drawing on insights from Australian urban water practitioners (n=90) across three Australian cities to explore the contemporary urban water governance context. The aim of this research is to provide guidance for shifting towards a water sensitive city (WSC) by showing that different components of governance strategies are more/less appropriate for achieving a WSC. The perceived effectiveness and 'fit' of current urban water governance strategies were explored by utilising a recently constructed fit-for-purpose governance framework. The fit-for-purpose governance framework helps assess whether the (anticipated) outcomes match the intended purposes of proposed and applied governance strategies. The research provides important insights regarding the need for a mix of centralised and decentralised, and formal and informal, governance approaches to support effective governance of water infrastructure operating across different scales. Thus, the different stages of transitioning to a water sensitive city will require different configurations of centralised/decentralised and formal/informal governance processes.

KEYWORDS
Governance, Transitions, Water sensitive city.

INTRODUCTION

Cities are highly dynamic. Moreover, they are subject to climate change and extreme weather. In particular, Australian cities are facing highly variable and extreme climate conditions. Over the last decade, long-lasting drought interrupted by short periods of extreme rainfall have put traditional, large-scale water infrastructure under pressure regarding the security of water supplies and protecting cities from flooding. In response to such pressures, the concept of a water sensitive city (WSC) has emerged concurrently from the technical and social science fields (Wong and Brown, 2009; Wong et al., 2011). A WSC is considered resilient to broadscale change (i.e. demographic change, climate change and extreme weather conditions) and values water, promotes conservation and aims to improve liveability. Such a city would achieve this through planning for diverse and flexible water sources (e.g. dams, desalination, water grids and stormwater harvesting), incorporating water sensitive urban design for flood mitigation, environmental protection and low carbon urban water services in the planning system, and enabling social and institutional capacity for sustainable water management (see also Wong and Brown, 2009; Wong et al., 2011). A WSC is the outcome of processes of Water Sensitive Urban Design (WSUD) (Wong et al., 2011).

World-wide, no WSC has been reported yet. However, creating WSCs is increasingly becoming a policy objective in Australia, with the terminology used by the National Water Commission and the South Australian Government (i.e. Adelaide's Water for Good strategy). Moreover, while technologies that make water sensitive cities possible have successfully been demonstrated on a number of occasions (Farrelly and Brown, 2011), there remain significant institutional barriers to facilitating this paradigm shift in planning, design, operation and management of urban water systems, including a lack of understanding about urban water cycles and different interpretations of water sensitive urban design (WSUD); the values of WSUD are not firmly embedded in the water and development sectors; limited skills and competencies to apply WSUD; fragmented urban water space; a limiting regulatory environment for technological innovation; and ineffective leadership (see also Brown and Farrelly, 2009; Maksimovic and Tejada-Guilbert, 2001; Pahl-Wostl, 2007). This paper explores ways to overcome these challenges and improve the effectiveness and 'fit' of new urban water governance strategies targeted at facilitating a water sensitive city. Because in particular improving stormwater management has been identified as a important factor to advance towards WSCs (Wong and Brown, 2009), we focus on this aspect.

THEORY

Transitioning to a WSC is a long term process (Brown et al., 2009). A transition is a structural change in the way a society or a subsystem of society (e.g. water management, energy supply, agriculture) operates, and can be described as a long-term non-linear process (25-50 years) that results from a co-evolution of cultural, institutional, economical, ecological and technological processes and developments on various scale levels (Rotmans et al., 2001). As such, transitions are structural changes of practices, institutions and culture. Managing transitions requires an adaptive governance approach, whereby there is continuous influence and adjustment in governance systems (Foxon et al., 2009; Smith and Stirling, 2010).

Governance is a concept that is defined and interpreted in many different ways (for an overview of definitions and interpretations, see e.g. Kjær, 2004; Rhodes, 1996). It refers to both processes and structures for steering and managing parts of societies (Kooiman, 1993; Pierre and Peters, 2000). Governance as process refers to managing networks, markets, hierarchies or communities (Kjær, 2004; Rhodes, 1996), whereas governance as structure refers to the institutional design of patterns and mechanisms in which social order is generated and reproduced (Voß, 2007). As such, governance relies on institutions consisting of cognitive (dominant knowledge, thinking and skills), normative (culture, values and leadership) and regulative components (administration, rules and systems) that mutually influence practice (Scott, 2001). Adaptive governance is a way of governing by anticipating long-term change (i.e. climate change, population growth), responding to immediate shock events (i.e. drought, flooding) and recovering from such events (see also Folke et al., 2005; Nelson et al., 2007).

To anticipate change and retain the ability to adjust strategies to changing drivers and problems, continuous monitoring, evaluation and adjustment of governance practices is required (Voß et al., 2006). However, the challenge to deal with complexity and uncertainty creates difficulties for many policy makers and practitioners to apply adaptive governance in practice (Rijke et al., submitted). In order to overcome these challenges, the fit-for-purpose governance framework is developed to evaluate the 'fit' of the urban water governance strategies (Rijke et al., submitted). This framework consists of three steps for diagnosing the fit-for-purpose of governance strategies: (1) identifying the purpose of governance; (2) mapping of the context; and (3) evaluating the outcome of governance mechanisms. As such, the fit-for-purpose of governance is an indication of the efficacy of governance to handle management issues for a particular purpose (i.e. securing water supplies).

METHODOLOGY

The aim of this research is to provide guidance for shifting towards a WSC by showing that different components of governance strategies are more/less appropriate for achieving a WSC. This research explicitly focuses upon exploring the strengths and weaknesses of existing governance strategies for transitioning towards a WSC in three Australian case study cities (Melbourne, Sydney and Adelaide). We focus on the responses that were implemented or planned in 2010; at the end of the drought that extended over a period of nearly a decade. To explore the strengths and weaknesses of the existing and emerging urban water governance strategies, the research has drawn upon the 'fit-for-purpose governance framework' recently developed by the authors (see Rijke et al., submitted). Applying this framework requires three steps: 1) developing an in-depth understanding of the historical and contemporary context of urban water governance within each of the three cities; 2) identifying existing and planned urban water governance strategies; and 3) assess to which extent the (anticipated) outcomes enable the establishment of water sensitive cities.

The case study research was shaped by Yin (2009) whereby a similar theoretical context for analysis in each of the three cities was used. Adelaide, Melbourne and Sydney were selected for each city has demonstrated recent innovation in stormwater harvesting predominantly driven by governments at different levels. For example, in Adelaide, local governments are taking up pioneering roles in setting up stormwater harvesting and reuse schemes, in Sydney local governments play a large role, while in Melbourne, leadership about stormwater harvesting is a combination of the water authority, state government and certain local governments Within each city, the network structures of practitioners, government representatives and community representatives, the ties between these actors, and changes in both over time were analysed in order to develop building blocks for effective strategies for stimulating transitions in urban water management.

Table B.0.1 **Overview of respondents in first interview round (July - September 2010)**

	Adelaide	Melbourne	Sydney	Total
State government	17	5	11	33
Local government	5	3	8	16
Water utilities	6	8	4	18
Professional associations	3	2	3	8
Private sector	2	4	3	9
Other	3	1	2	6
Total	**36**	**23**	**31**	**90**

To provide an in-depth reflection upon existing and planned strategies for urban water governance in each city, leading urban water practitioners (n=90; see Table 1) were involved in semi-structured interviews focused upon the (historical) context of urban water governance, activities to achieve a WSC, and the strengths and limitations of these activities. Interviewees represented a range of different disciplines and organisations, which included key decision makers and individuals in senior advisory roles. Furthermore, critical policy and media documentation were also collated to support and/or contradict practitioner interpretations. Following the preliminary data analysis, a results validation process was undertaken. This involved i) a series of validation workshops (May-June 2011) with participants (n=81) in each city and ii) interviews (n=15) with 20 representatives of key stakeholder groups which were previously interviewed.

URBAN WATER GOVERNANCE IN AUSTRALIA

Traditionally, Australian urban water management is largely the responsibility of State Governments, which have predominantly relied on a highly centralised 'big pipes-in big pipes-out' infrastructure (see, e.g. Newman, 2001). Each of the cities has its own specific urban water context: For example, in Adelaide, local governments are playing a strong leadership role since the early 1990s with regards to innovation of stormwater harvesting and reuse technologies. For example, in Melbourne, has a long history with demonstration of stormwater treatment technologies and has a mandate for WSUD in its State planning regulation. For example, Sydney has one of the largest per capita water storage systems in the world and New South Wales Government was the first State Government in Australia to enable urban water markets through the introduction of the Water Industry Competition Act 2006.

However, in recent years, all three of the case study cities have faced similar macro-scale drivers for reconsidering conventional urban water management practices. For example, a prolonged drought significantly affected the three cities for almost a decade (2001-2009). This situation forced state governments to restrict residential water use, which in turn made water management a key political issue with widespread media attention. Consequently, to secure water supplies, each city has augmented its water supply resources by constructing rainfall-independent desalination plants and waste water recycling schemes. In addition, investments have been made in supporting rainwater tanks and stormwater harvesting and reuse schemes. At a national scale, the Government has established the 'National Water Initiative' (NWI) with the purpose of undertaking widespread efficiency and effectiveness reforms in rural and urban water sectors. Using a series of key water supply, efficiency and pricing innovations, the NWI aims to achieve a reliable, healthy, safe and sustainable urban water supply and more liveable, sustainable and economically prosperous cities (National Water Commission, 2009, 2011).

RESULTS AND DISCUSSION

Based on the research data and the findings of two previous research projects (2004-2011), we have identified a number of dominant governance structures that enable sustainable urban water management practice and a number of processes that are needed to transform towards a water sensitive city state (see Farrelly et al., submitted). The research data show that these are centralised to different degrees, depending on the scale level of infrastructure systems, involvement of federal, state, regional or local government agencies (see Table 2). Furthermore, as Figure 1 shows, we conclude from our research data that urban water governance relies on a mix of formal institutions and informal networks. Based on the research findings, we conclude that urban water governance in the three cities is, albeit to different degrees, a mostly centralised affair. In Adelaide, governance is, to a large extent, coordinated by the Department for Water that was established in 2010 as the leading urban water policy organisation. A strong informal network of planners, engineers and policy makers across all stakeholder groups provides, with fluctuating intensity, input to governance processes. In Melbourne, a similar, but slightly less connected, informal network was identified. However, compared to Adelaide, there is less centralised coordination of urban water governance. Instead, leadership is distributed more amongst key stakeholder groups, such as Melbourne Water, water retailers, and local governments and research institutions. In Sydney, governance relies almost exclusively on formal institutions. Informal networks were identified, but they mostly operate within formalised coalitions of local governments around stormwater management.

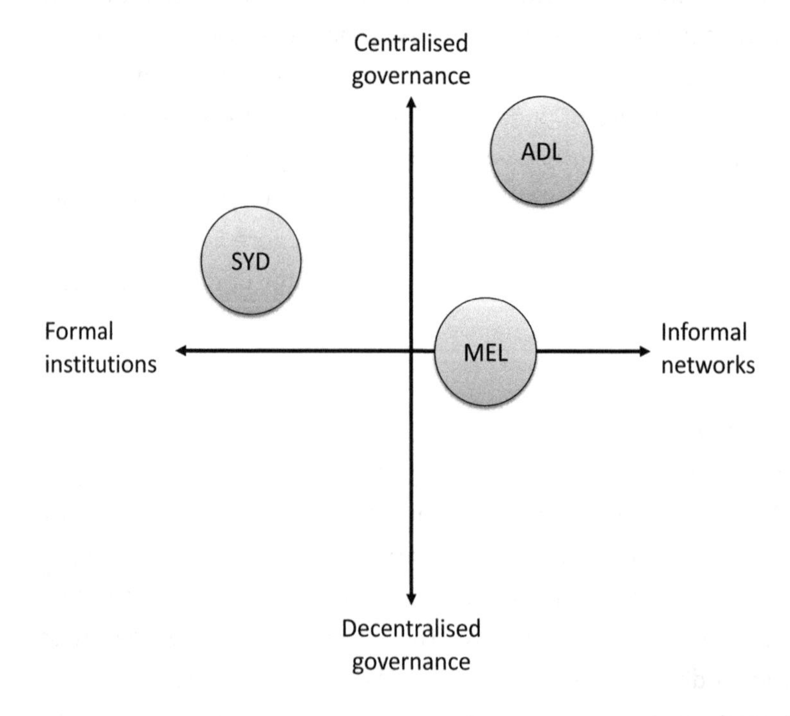

Figure B.1 **Urban water governance in Adelaide (ADL), Melbourne (MEL) and Sydney (SYD)**

To get an indication of the 'fit' of the approaches that are applied in these cities, we have identified the strengths and weaknesses based on scientific literature, respondents' perceptions, and the validation. In order to keep this paper comprehensive, we have selected four aspects from Table 1 and summarised their strengths and weaknesses below (to avoid repetition of arguments below, we merged 'Leadership' and 'Policy framework and institutional design'):

Leadership and policy making: Centralised efforts may lead to effective coordination of activities to create synergy between projects and use resources (time/budget) efficiently. However, centralised approaches are often unsuccessful in solving complex problems and may cause others to walk away from responsibilities. Decentralised (collaborative) leadership and policy making occurs often in conjunction of learning alliances that provide "safe spaces". Therefore, they reduce the risk of illegitimate, unequal and/or unfair outcomes. However, decentralised approaches may result in less effective coordination and use of resources.

Regulation and compliance: Centralised state-wide regulation creates an new bottom line and a level playing field for all stakeholders. For example, respondents comment that a mandate for WSUD removes inconsistencies in development processes. However, sufficient awareness about value, capacity, competences and guidelines is needed to implement regulation in practice. It was identified that establishing state-wide regulation, for example for WSUD, depends on political leadership. Decentralised regulation of WSUD enables tailoring to local contexts and provides an opportunity to overcome lack of centralised action. However, barriers need to be overcome multiple times in different (local government) organisations.

Capacity building and demonstration: Central coordination of capacity building efforts enables greater knowledge sharing within and across organisations through collecting, collating, synthesising and distributing project outcomes and lessons learned. However, this is resource intensive. Decentralised capacity building efforts, for example through learning alliances, were also identified. Often, these alliances aim to advocate alternative practices. Sometimes they build on multiple sources of knowledge, but a risk of limited knowledge sharing also came apparent. Formalised coalitions provided opportunities for informal collaboration and building of trust.

Table B.2 Overview of governance structures and processes in Adelaide (ADL), Melbourne (MEL) and Sydney (SYD).

STRUCTURES	**Centralised**	**Decentralised**
Vision (narrative, metaphor, images)	In ADL and MEL, the language of water sensitive cities is not mainstream yet, but it is increasingly being adopted by the main stakeholders. E.g. it is used in the national Cities of the Future platform that originated in MEL and in ADL's urban water policy *Water for Good.*	No joint activities for adopting a vision for water sensitive cities in SYD: whilst State Government focuses on water security, local governments focus more on waterway health and urban drainage.
Policy framework and institutional design	*Water for Good* (ADL) and the *2010 Metropolitan Water Strategy* (SYD) are both a first step towards integrated urban water management. However, these strategies primarily on water security and to a lesser extent on water quality, but do not incorporate flooding.	MEL's State Government policy framework for WSUD is catching up with practice. However, currently policy arrangements are distributed over a number of organisations such as the water authority, water retailers, and several local governments.
Economic incentives and justification	Several projects in ADL and MEL were co-funded by a $200 million fund of the Federal Government for stormwater harvesting and reuse projects. The eight projects in ADL are centrally coordinated through a Stormwater Programme at the Department for Water.	SYD's regulation enables competition on the urban water market and thrid party access of water infrastructure to achieve economic efficiency and diversification of water supply. Also in ADL and, to a lesser extent MEL, an ambition to establish urban water markets was identified.
Regulation and compliance	Since 2006, WSUD is mandated for new residential developments in Victoria (MEL). South Australia (ADL) has announced a similar state-wide mandate. Also public health and environmental protection are regulated through state-wide regulation (all cities).	Rather than a state-wide mandate, activities are undertaken to institutionalise WSUD through local governments' planning regulation (LEPs), so that planners and developers in councils with such regulation have an obligation to consider the implementation of WSUD.
PROCESSES		
Leadership	Through ministerial decisions and formal institutional mandates and responsibilities. (all cities, particularly in SYD and ADL).	Through active collaboration in informal networks (ADL, MEL) and through collaboration in demonstration projects and learning alliances (all cities).
Capacity building and demonstration	Capacity building organised by Clearwater in MEL. The need for a similar program is identified amongst all stakeholder groups in ADL. The 'WSUD in Sydney' programme acts as a knowledge broker and network manager (SYD).	ADL is pioneering with Aquifer Storage and Recovery since the early 90s. MEL is demonstrating stormwater treatment technologies since the late 90s. Local alliances in SYD are established to learn about and advocate for implementation of WSUD schemes.
Public en-	Awareness programmes, education,	Project-based engagement for cre-

gagement	water restrictions and rain-tank programmes (all cities)	ating public support and public participation and co-management for identifying problems and solutions (all cities)
Research partnerships with policy and practice	De Goyder Institute (ADL), Centre for Water Sensitive Cities (MEL/national) and Cooperative Research Centres (national) combine and coordinate research capacities of various institutes and form partnerships with policy and practice.	Uncoordinated collaborative research of universities and research institutions in policy evaluations and advice, capacity building projects, and demonstration projects (all cities)

Similarly, we have analysed the strengths and weaknesses of formal and informal modes of governance. Traditionally, governance relies on formal institutions in which legislative frameworks set out the rules and responsibilities of involved organisations. Formal agreements are often binding and difficult to reverse. However, our data showed that this invokes a silo mentality between different departments within and between organisations and results in low levels of adaptive capacity. It was found that informal networks of practitioners and decision makers play an important role in building trust between disciplines and organisations. Because they operate outside the scrutiny of organisational mandates, it was found that they are a consistent driver for exploring and understanding problems and solutions and a starting point and incubator for innovative ideas. The identified informal networks are highly adaptive and contain high degrees of tacit knowledge. They are vulnerable to losing such knowledge, because they operate on a voluntary basis and are thus prone to 'job-hopping' of network members and fluctuating levels of interactions. A large number of respondents identified a need for improving methods for formal institutions to tap from informal networks without losing the ability of unscrutinised behaviour of their members.

In order to establish water sensitive cities, urban water governance needs to be both adaptive and transformative: it needs to be adaptive in order to deal with ever changing circumstances and purposes of governance, and it needs to be transformative to enable a transition from current urban water practices towards water sensitive cities. Adaptive governance is a matter of continuous learning and making timely decisions. Depending on the circumstances, a different mix of centralised, decentralised, formal and informal approaches can be considered more favourable. For example, in times of immediate crisis, such as the threat of running out of water supplies that occurred in Adelaide in 2006, the adaptive strength of informal networks could be utilised to mobilise and inform quick centralised coordination of responses. Whereas, more decentralised approaches may be preferred to challenge existing dominant regimes or to create legitimate and fair long term policy frameworks. Transformative governance, also referred to as transition management, aims to enable progress towards a water sensitive city. Different stages of the transition process towards water sensitive cities favour different configurations of centralised and decentralised governance processes. For example, the predevelopment stage of a transition typically involves the invention of new technologies and processes and the formation of informal networks. As shown above, decentralised and informal governance approaches are favourable to enable such activities. On the other hand, a transition approaches completion when a new set of institutional arrangements to safeguard a new status quo is established; often the result of a centralised and formal governance approach.

CONCLUSION

In this paper, the strengths and weaknesses of contemporary urban water governance approaches are identified. We conclude that a mix of centralised and decentralised governance approaches is needed for effective governance of water infrastructure that operates across different scales. Also we showed that formal institutions need to be complemented with informal networks in which innovative ideas are discussed and developed. Upon reflection, the fit for purpose framework has been a useful analytical tool to examine contemporary urban water governance approaches and the assessment undertaken in this research provide guidance for: 1) developing effective adaptation strategies, and 2) speeding up a transition towards water sensitive cities.

ACKNOWLEDGEMENT

This research was made possible by the Cities as Water Supply Catchments research programme (www.watersensitivecities.org.au) that is funded by a consortium of public and private organisations. The authors thank the funders for their support and encouragement to undertake this research.

REFERENCES

Brown, R., Farrelly, M., 2009. Delivering sustainable urban water management: a review of the hurdles we face. Water science and technology: a journal of the International Association on Water Pollution Research 59, 839.

Brown, R., Keath, N., Wong, T., 2009. Urban water management in cities: historical, current and future regimes. Water science and technology 59, 847-855.

Farrelly, M., Brown, R., 2011. Rethinking urban water management: Experimentation as a way forward? Global Environmental Change 21, 721-732.

Farrelly, M., Rijke, J., Brown, R., submitted. Exploring the key ingredients in enabling transitions governance, 7th International Conference on Water Sensitive Urban Design, Melbourne, Australia.

Folke, C., Hahn, T., Olsson, P., Norberg, J., 2005. Adaptive governance of social-ecological systems. Annual Review of Environment and Resources 30, 441.

Foxon, T.J., Reed, M.S., Stringer, L.C., 2009. Governing long term social–ecological change: what can the adaptive management and transition management approaches learn from each other? Environmental Policy and Governance 19, 3-20.

Kjær, A., 2004. Governance, Cambridge: Polity.

Kooiman, J., 1993. Modern governance: new government-society interactions. Sage Publications Ltd.

Maksimovic, C., Tejada-Guilbert, J., 2001. Frontiers in urban water management: Deadlock or hope. Intl Water Assn.

National Water Commission, 2009. Australian Water Reform 2009: Second biennial assessment of progress in implementation of the National Water Initiative, In: NWC (Ed.), Canberra, ACT, Australia.

National Water Commission, 2011. Urban water in Australia: future directions, In: NWC (Ed.), Canberra, ACT, Australia.

Nelson, D.R., Adger, W.N., Brown, K., 2007. Adaptation to environmental change: contributions of a resilience framework. Annual Review of Environment and Resources 32, 395.

Pahl-Wostl, C., 2007. Transitions towards adaptive management of water facing climate and global change. Integrated Assessment of Water Resources and Global Change, 49-62.

Pierre, J., Peters, B., 2000. Governance, politics and the state. St. Martin's Press.

Rhodes, R., 1996. The New Governance: Governing without Government. Political studies 44, 652-667.

Rijke, J., Brown, R., Zevenbergen, C., Ashley, R., Farrelly, M., Van Herk, S., Morison, P., submitted. Fit-for-purpose governance: A framework to operationalise adaptive governance.

Rotmans, J., Kemp, R., van Asselt, M., 2001. More evolution than revolution: transition management in public policy. Foresight 3, 15-31.

Scott, W.R., 2001. Institutions and organizations. Sage Publications, Inc.

Smith, A., Stirling, A., 2010. The politics of social-ecological resilience and sustainable socio-technical transitions. Ecology and Society 15, 11.

Voß, J.P., 2007. Innovation processes in governance: the development of'emissions trading'as a new policy instrument. Science and Public Policy 34, 329-343.

Voß, J.P., Bauknecht, D., Kemp, R., 2006. Reflexive governance for sustainable development. Edward Elgar Publishing.

Wong, T., Brown, R., 2009. The water sensitive city: principles for practice. Water science and technology 60, 673-682.

Wong, T.H.F., Allen, R., Beringer, J., Brown, R.R., Chaudhri, V., Deletic, A., Fletcher, T.D., Gernjak, W., Hodyl, L., Jakob, C., Reeder, M., Tapper, N., Walsh, C., 2011. blueprint2011: Stormwater Management in a Water Sensitive City Centre for Water Sensitive Cities,, Melbourne.

Acknowledgements

Thinking back to my time as a PhD candidate makes me smile. It was mostly fun, sometimes challenging, almost always insightful and it added meaning to my life. Making up the checks and balances, I can conclude that I've experienced many things, learned a lot and transformed along the way. In particular, I was fortunate to interact with several people who inspired and supported me. You had a big role in making this period the time of my life!

Chris, thank you for being my promotor in the broadest sense of the word. You inspired, stimulated and challenged me as a researcher, professional and a person. Often with seemingly simple question, a peculiar sensitivity to relevant matters and always with a great dose of empathy. You gave me freedom to pursue the opportunities that came on my path and created excellent conditions for me to flourish. Thank you for talking me into writing a PhD thesis!

Rebekah, thank you for giving me the opportunity to come back to Monash as a PhD researcher and giving me the trust to explore the networks of key people of the Australian urban water sector as one of the earliest projects of the Cities as Water Supply Catchments programme. Moreover, thank you for your excellent guidance on the methodological and theoretical aspects of my thesis. Each of our Skype calls or personal meetings has brought my thinking further.

Richard, I feel honoured that you shared your experience with me and provided all my work of critical reality checks. Your humorous comments made revising papers significantly more enjoyable. However, they also warned me that I couldn't get away with cutting corners or imprecise formulations to masquerade unfinished lines of thought. And because of the interest you show in my family, I let you get away with calling me a schweinhund every now and then.

Sebastiaan, buddy, we have been a winning team. Our interviews would make even Pauw and Witteman jealous. The discussions about the content of our work, the way we should go about it and how to get the most out of it really paid off. Your unsolicited advice is always highly appreciated. Our collaboration shows how successful peer pressure is when it is applied between friends that wish the best for each other.

FRG buddies, thank you for your outspokenness, straight-forwardness, transparency, bluntness, humour and loyalty. I'll never forget our joint travels to, for example, Melbourne, Porto Alegre, Istanbul, New Orleans and Barcelona. Special thanks to William, who helped me overcoming extreme cash-flow problems caused by unwilling bureaucrats, and Ellen, who helped out with the organisation of many interviews.

Megan, thank you for jumping into a running project about a complex subject and analysing data that you didn't collect yourself. You were a fantastic super-sub! Having become a father myself now, I fully understand how challenging it was to have weekly Skype calls right after "the witching hour".

Peter, thank you for leaving your family a few weeks to conduct interviews with me. Our average of 15 interviews per week would not have been possible by myself and not half as enjoyable. It was amazing to learn from all the personal stories about how to establish change. We certainly "got some f-in' runs on the board"!

Thank you Ingwer, Liesbeth and Ben, for showing me around in the Room for the River family as its father, mother and wise uncle. Listening and discussing with the people in the programme learnt me many things about programme management, collaboration and doing things that have not been done before.

Mum and dad, this thesis is proof that your education has worked. I pursued things that I liked and persisted in doing them as good as I could, even when other things distracted me. I have to confess that it was not always easy to stay focused, but completion gave me a proud and satisfied feeling. Thank you for teaching me this.

Son, at the start of my PhD, you were the beautiful girl next door with whom I fell in love. At present, you are still beautiful and we have become the mother and father of our fantastic son Bram. In between, you have endured my absence when I was travelling and bad moods when I suffered from writer's block or ordinary PhD frustration. Thank you for always being there for me. I love you.

About the author

Jeroen Rijke is a consultant at Triple Bridge and a researcher at UNESCO-IHE. His work focuses on assisting policy makers, planners and project managers with developing adaptive strategies and management approaches for sustainable infrastructure, mainly in the water sector. Jeroen is a civil engineer (TU Delft) who is trained in policy sciences (KTH, Monash University) and has worked across Europe, Australia and Africa.

His PhD focused on the governance of adaptation to flooding and drought (TU Delft and UNESCO-IHE). He has analysed governance responses to a decade of drought in Sydney, Melbourne and Adelaide to provide building blocks for enhancing innovative urban stormwater management practices and establishing transitions towards 'water sensitive cities'. In addition, he evaluated multi-level governance processes within the €2.4 billion Room for the River programme in the Netherlands to develop recommendations for Rijkswaterstaat and the Delta Programme to improve existing programme management practices and establish adaptive delta management.

Prior to his PhD research, Jeroen mainly focused on climate adaptation in urban areas. He was lead author of a State of the Art report about climate adaptation in urban areas that played an instrumental role in the programming of the research on this topic of the Dutch national Knowledge for Climate research programme. Also, he was involved in a national assessment of the vulnerability and adaptive capacity of Dutch cities to climate change that was conducted for the Netherlands Environmental Assessment Agency (PBL).

Jeroen is a frequent reviewer of manuscripts submitted to international conferences and international scientific journals, including, for example, the Journal of Water and Climate Change, Natural Hazards and the Journal of Flood Risk Management. In addition, he is currently acting as a guest-editor for a special issue on making space for rivers in the International Journal of Water Governance.

Other publications by the author

Peer reviewed journal articles and book chapters

1. Rijke, J., van Herk, S., Zevenbergen, C., Ashley, R., Hertogh, M., ten Heuvelhof, E. (in press) Adaptive programme management through a balanced performance/strategy oriented focus. International Journal of Project Management.
2. Salinas Rodriguez, C., Ashley, R., Gersonius, B., Rijke, J., Pathirana, A., Zevenbergen, C. (in press) Incorporation and application of resilience in the context of water sensitive urban design: Linking European and Australian perspectives. WIREs Water.
3. van Herk, S. Rijke, J., Zevenbergen, C., Ashley, R. Besseling, B. (in press) Adaptive co-management and network learning in the Room for the River programme. Journal of Environmental Planning and Management.
4. Rijke, J., Farrelly, M., Brown, R., Zevenbergen, C. (2013) Configuring transformative governance to enhance resilient urban water systems. Environmental Science and Policy 25: 62-72.
5. van Herk, S., Rijke, J., Zevenbergen, C., Ashley, R., (2013) Understanding the transition to integrated flood risk management in the Netherlands. Journal of Environmental Innovations and Societal Transitions.
6. Rijke. J., Brown, R., Zevenbergen, C., Ashley, R., Farrelly, M., Morison, P. and van Herk, S. (2012) Fit-for-purpose governance: a framework to make adaptive governance operational. Environmental Science & Policy, 22: 73-84.
7. Bettini, Y., Rijke, J., Farrelly, M. & Brown, R. R. (2013) Connecting levels and disciplines: Connective Capacity of Institutions and Actors Explored. Chapter 7 in: Edelenbos, J., Bressers, N. & Scholten, P. (Eds.) Water Governance as Connective Capacity. Ashgate. ISBN: 978-1-4094-4746-7. eISBN 978-1-4094-4747-4 (Published January 2013).
8. Rijke, J., van Herk, S., Zevenbergen, C., Ashley, R. (2012) Room for the River: Delivering integrated river basin management in the Netherlands. International Journal of River Basin Management 10(4): 369-382.
9. Zevenbergen, C., van Herk, S., Rijke, J., Kabat, P., Bloemen, P., Speers, A., Gersonius, B., Veerbeek, W. (2012) Taming global flood disasters. Lessons learned from Dutch experience. Natural Hazards 1-9.
10. van Herk, S., Zevenbergen, C., Ashley, R., Rijke, J. (2011) Learning and Action Alliances for the integration of flood risk management into urban planning: a new framework from empirical evidence from The Netherlands. Environmental Science & Policy 14, 543-554.

11. van Herk, S., Zevenbergen, C., Rijke, J., Ashley, R. (2011) Collaborative research to support transition towards integrating flood risk management in urban development. Journal of Flood Risk Management 4, 306-317.

Conference proceedings

12. Rijke, J., Vessels Smith, J., Pathirana, A., Gersonius, B., Ashley. R., Wong, T., Zevenbergen, C. (2013) Three layers of intervention for achieving drought resilience: protection, prevention, preparedness. 8th International Water Sensitive Urban Design Conference, 25-29 November 2013, Gold Coast, Australia.
13. Salinas Rodriguez, C., Ashley, R., Gersonius, B., Pathirana, A., Rijke, J., Zevenbergen, C. (2013) Flood resilience incorporation and application in the context of water sensitive cities. 8th International Water Sensitive Urban Design Conference, 25-29 November 2013, Gold Coast, Australia.
14. van Herk, S. Rijke, J., Zevenbergen, C., Ashley, R., Besseling, B. (2013) Adaptive multi-level governance through social learning: River Basin Management in the Netherlands. Earth System Governance Tokyo Conference, 28-31 January 2013, Tokyo, Japan.
15. Rijke, J., van Herk, S., Zevenbergen, C., Ashley, R., (2012) Towards integrated river basin management: governance lessons from Room for the River, Floodrisk 2012 - 2nd European Conference on flood risk management, Rotterdam, Netherlands.
16. Rijke, J., van Herk, S., Zevenbergen, C., Ashley, R. (2012) A programme management approach for governing a transition to integrated flood management in the Netherlands. International Conference on Sustainability Transitions, Copenhagen, Denmark.
17. Rijke, J., Farrely, M., Brown, R., Zevenbergen, C. (2012) Creating water sensitive cities in Australia: the strengths and weaknesses of current governance approaches. 7th International Conference on Water Sensitive Urban Design, Melbourne, Australia.
18. Farrelly, M., Rijke, J., Brown, R. (2012) Exploring operational attributes of governance for change, 7th International Conference on Water Sensitive Urban Design, Melbourne, Australia.
19. van Herk, S., Rijke, J., Zevenbergen, C., Ashley, R. (2012) Transition in governance of River Basin Management in the Netherlands through social learning. International Conference on Sustainability Transitions, Copenhagen, Denmark.

20. van Herk, S., Rijke, J., Zevenbergen, C., Ashley, R., (2012) Governance of integrated flood risk management to deliver large scale investment programmes: delivery focused social learning in the Netherlands., Floodrisk 2012 - 2nd European Conference on flood risk management, Rotterdam, Netherlands.
21. Veerbeek, W., Ashley, R., Zevenbergen, C., Gersonius, B., Rijke, J. (2012) Building adaptive capacity for flood proofing in urban areas through synergistic interventions. 7th International Conference on Water Sensitive Urban Design, Melbourne, Australia.
22. Rijke, J., Farrelly, M., Morison, P., Brown, R., Zevenbergen, C., (2011) Towards improved urban water governance in Adelaide, Australia, 12th International Conference on Urban Drainage, Porto Alegre, Brazil.
23. Rijke, J., Anema, K., Zevenbergen, C., van de Ven, F., van Herk, S., (2011) Capacity building dissected: what to learn in order to mainstream sustainable urban water management in the Netherlands?, 2nd International Conference on Sustainability Transitions, Lund, Sweden.
24. Anema, K. and Rijke, J. (2011) Putting new climate adaptation measures into practice: why bother? Resilient Cities 2011 - 2nd World congress on cities and adaptation to climate change, Bonn, Germany.
25. Rijke, J., Veerbeek, W., Zevenbergen, C., van Herk, S. van de Ven, F. (2010) Adapting where we can, instead of where we have to. International Conference Deltas in Time of Climate Change, 29 September - 1 October 2010, Rotterdam, The Netherlands.
26. Rijke, J., Morison, P., Brown, R. and Zevenbergen, C. (2010) Framing Risk perceptions in transitions to drought and flood resilient urban development. International conference Deltas in Time of Climate Change, 29 September - 1 October 2010, Rotterdam, The Netherlands.
27. Bax, J., Rijke, J., van der Meulen, F., Seck, A., Gaye, M., Waals, H., Kelder, E., Zevenbergen, C. (2010) Learning together to manage urban flood risks in Dordrecht and Saint Louis. International Conference Deltas in Times of Climate Change, 29 september - 1 October 2010, Rotterdam, The Netherlands.
28. Veerbeek, W., Ashley, R., Zevenbergen, C., Rijke, J. S. and Gersonius, B. (2010) Building adaptive capacity for flood proofing in urban areas through synergistic interventions, in: Proceedings of the ICSU 2010 First International Conference on Sustainable Urbanization, Hong Kong Polytechnic University, Faculty of Construction and Land Use, Hong Kong.
29. van Herk, S., Zevenbergen, C., Rijke, J., Ashley, R. (2009) Collaborative research to support transition towards integrating flood risk management in urban development, International Conference on Urban Flood Management, Paris, France, 26-27 November 2009.

30. Rijke J.S., De Graaf R.E., Van de Ven F.H.M., Brown R.R. and Biron D.J. (2008) Comparative case studies towards mainstreaming water sensitive urban design in Australia and the Netherlands. 11th International Conference on Urban Drainage, Edinburgh, Scotland, August 31–September 5, 2008.

Other publications

31. Zevenbergen, C., Rijke, J., van Herk, S., Ludy, J., Ashley, R. (2013) Room for the River: international relevance. Water Governance 2: 24-31.
32. Rijke, J. Various sections in: Zevenbergen,C., Cashman, A., Evelpidou, N., Pasche, E., Garvin, S., Ashley, R. (eds) (2010) Urban Flood Management. ISBN: 978-0-415-55944-7, CRC Press.
33. van der Ven, F., Van Nieuwkerk, E., Stone, K., Veerbeek, W., Rijke, J., van Herk, S., Zevenbergen, C. (2010) Building the Netherlands Climate Proof: Urban areas, Technical Report 1201082-000-VEB-0003, Deltares and UNESCO-IHE, Utrecht/Delft.
34. Rijke, J., Zevenbergen, C. and Veerbeek, W. (2009). State of the art Klimaat in de Stad, ISBN 978-94-90070-07-6, KvK rapportnummer KvK007/2009
35. Rijke J.S. (2007) Mainstreaming innovations in urban water management - Case studies in Melbourne and the Netherlands. MSc thesis, Delft University of Technology, Delft, Netherlands.

PGMO 06/27/2018